BARRON'S

SAT® SUBJECT TEST

MATH LEVEL 2

2008

8TH EDITION

Richard Ku, M.A.
AP Math Teacher and Former
Math Department Head
North Kingstown High School
North Kingstown, Rhode Island

and

Howard P. Dodge, M.A.
Former Math Teacher
The Wheeler School
Providence, Rhode Island

BARRON'S

About the Authors

Richard Ku has been teaching secondary mathematics, including Algebra 1 and 2, Geometry, Precalculus, AP Calculus, and AP Statistics, for over 25 years. He has coached math teams for 15 years and has also read AP Calculus exams for five years and began reading AP Statistics exams in 2007.

Howard P. Dodge spent 40 years teaching math in independent schools before retiring.

Acknowledgments

Thanks to my wife, Debra, for her support, encouragement, patience, and understanding while I worked evenings and weekends during the past year. My thanks also to Howard Dodge for asking me to revise the 7th and now the 8th editions of the book he began writing years ago. I would also like to thank Barron's editor Pat Hunter for guiding me through the preparation of this new edition.

© Copyright 2008 by Barron's Educational Series, Inc. Previous edition © Copyright 2003, 1998 under the title *How to Prepare for the SAT II: Math Level IIC*. Prior editions © Copyright 1994 under the title *How to Prepare for the SAT II: Mathematics Level IIC* and © Copyright 1991, 1987, 1984, 1979 under the title *How to Prepare for the College Board Achievement Test—Math Level II* by Barron's Educational Series, Inc.

All inquiries should be addressed to:
Barron's Educational Series, Inc.
250 Wireless Boulevard
Hauppauge, New York 11788
www.barronseduc.com

ISBN-13: 978-0-7641-3692-4 (book only)
ISBN-10: 0-7641-3692-5 (book only)
ISBN-13: 978-0-7641-9345-3 (book/CD-ROM package)
ISBN-10: 0-7641-9345-7 (book/CD-ROM package)

ISSN 1939-3237 (book only)
ISSN 1939-3245 (book/CD-ROM package)

PRINTED IN THE UNITED STATES OF AMERICA

9 8 7 6 5 4 3 2 1

Contents

GRAPHING CALCULATORS

MODEL TESTS

APPENDIX

Introduction

The purpose of this book is to help you prepare for the SAT Level 2 Mathematics Subject Test. This book can be used as a self-study guide or as a textbook in a test preparation course. It is a self-contained resource for those who want to achieve their best possible score.

Because the SAT Subject Tests cover specific content, they should be taken as soon as possible after completing the necessary course(s). This means that you should register for the Level 2 Mathematics Subject Test in June after you complete a precalculus course.

You can register for SAT Subject Tests at the College Board's web site, *www.collegeboard.com*; by calling (866) 756-7346, if you previously registered for an SAT Reasoning Test or Subject Test; or by completing registration forms in the SAT Registration Booklet which can be obtained in your high school guidance office. You can also write to:

College Board SAT Program
P.O. Box 025505
Miami, FL 33102

You may register for up to three Subject Tests at each sitting.

Colleges use SAT Subject Tests to help them make both admission and placement decisions. Because the Subject Tests are not tied to specific curricula, grading procedures, or instructional methods, they provide uniform measures of achievement in various subject areas. This way, colleges can use Subject Test results to compare the achievement of students who come from varying backgrounds and schools.

You can consult college catalogs and web sites to determine which, if any, SAT Subject Tests are required as part of an admissions package. Many "competitive" colleges require the Level 1 Mathematics Test.

If you intend to apply for admission to a college program in mathematics, science, or engineering, you may be required to take the Level 2 Mathematics Subject Test. If you have been generally successful in high school mathematics courses and want to showcase your achievement, you may want to take the Level 2 Subject Test and send your scores to colleges you are interested in even if it isn't required.

NEW IN THE 8TH EDITION

The 8th edition is the result of revisions that reflect both the organization and content of the current Level 2 Subject Test. The Review of Major Topics in Part 2 is now grouped into four chapters:

- Functions
- Geometry and Measurement
- Numbers and Operations
- Data Analysis, Statistics, and Probability

These correspond to the four content areas identified by the College Board for both the Level 1 and Level 2 Subject Tests.

Some content that was in earlier editions of this book has been deleted. These include:

- The relation between the zeros and coefficients of cubic and higher-degree polynomials
- Trigonometric formulas for the sum and difference of angles, and half-angle formulas
- Trigonometric form of complex numbers

- DeMoivre's theorem
- Three-dimensional coordinate geometry concepts of trace, direction numbers, and direction cosines

Other content has been added or enhanced:

- Transformations and symmetry
- Counting (e.g., permutations and combinations)
- Complex numbers
- Matrices
- Recursive sequences
- Statistical concepts of standard deviation and linear least squares regression

Also new to the 8th edition are two model tests on CD-ROM that appear in books with a CD-ROM.

STRUCTURE OF THIS BOOK

The overall structure of the 8th edition remains essentially the same as its predecessor. A Diagnostic Test in Part 1 follows this introduction. This test will help you quickly identify your weaknesses and gaps in your knowledge of the topics. You should take it under test conditions (in 1 quiet hour), complete the self-evaluation chart at the end, and read the solution explanations for the problems that you got wrong. These explanations include a code for calculator use (which is explained later), the correct answer choice, and the location of the relevant topic in the Part 2 "Review of Major Topics." For your convenience, a self-evaluation chart is also keyed to these locations.

Instructional material on the use of graphing calculators can be found in Part 3. The majority of those taking the Level 2 Mathematics Subject Test are accustomed to using graphing calculators. Where appropriate, explanations of problem solutions are based on their use. (Secondary explanations rely on algebraic techniques, where possible.) The material in Part 3 is designed to fill gaps in the graphing calculator experience of Level 2 test takers.

Part 4 contains six model tests. The breakdown of test items by topic approximately reflects the nominal distribution established by the College Board. The percentage of questions for which calculators are required or useful on the model tests is also approximately the same as that specified by the College Board. The model tests are self-contained. Each has an answer sheet and a complete set of directions. Each test is followed by an answer key, explanations such as those found in the Diagnostic Test, and a self-evaluation chart.

OVERVIEW OF THE LEVEL 2 SUBJECT TEST

The SAT Mathematics Level 2 Subject Test is one hour in length and consists of 50 multiple-choice questions, each with five answer choices. The test is aimed at students who have had two years of algebra, one year of geometry, and one year of trigonometry and elementary functions. According to the College Board, test items are distributed over topics as follows:

- Numbers and Operation: 5–7 questions
  Operations, ratio and proportion, complex numbers, counting, elementary number theory, matrices, sequences, series, and vectors

- Algebra and Functions: 24–26 questions
  Work with equations, inequalities, and expressions; know properties of the following classes of functions: linear, polynomial, rational, exponential, logarithmic, trigonometric and inverse trigonometric, periodic, piecewise, recursive, and parametric

- Coordinate Geometry: 5–7 questions
  Symmetry, transformations, conic sections, polar coordinates

- Three-dimensional Geometry: 2–3 questions
  Volume and surface area of solids (prisms, cylinders, pyramids, cones, and spheres); coordinates in 3 dimensions

- Trigonometry: 6–8 questions
  Radian measure; laws of sines and law of cosines; Pythagorean theorem, cofunction, and double-angle identities

- Data Analysis, Statistics, and Probability: 3–5 questions
  Measures of central tendency and spread; graphs and plots; least squares regression (linear, quadratic, and exponential); probability

CALCULATOR USE

As noted earlier, most taking the Level 2 Mathematics Subject Test will use a graphing calculator. In addition to performing the calculations of a scientific calculator, graphing calculators can be used to analyze graphs and to find zeros, points of intersection of graphs, and maxima and minima of functions. Graphing calculators can also be used to find numerical solutions to equations, generate tables of function values, evaluate statistics, and find regression equations. Although a graphing calculator may provide an advantage on certain questions, you should use the calculator with which you are most comfortable.

You should always read a question carefully and decide on a strategy to answer it before deciding whether a calculator is necessary. You may find that you need a calculator only to evaluate some expression that must be determined based solely on your knowledge about how to solve the problem.

Graphing calculators and newer scientific calculators are very user friendly. Because they follow order of operations and expressions can be entered using several levels of parentheses, there is never a need to round and write down the result of an intermediate calculation and then rekey that value as part of another calculation. Premature rounding can result in choosing a wrong answer if the answer choices are close in value.

On the other hand, graphing calculators can be troublesome or even misleading. For example, if you have difficulty finding a window for a graph, perhaps there is a better way to solve a problem. Piecewise functions, functions with restricted domains, and functions having asymptotes provide other examples where the usefulness of a graphing calculator may be limited.

Calculators have popularized a multiple-choice problem-solving technique called backsolving, where answer choices are entered into the problem to see which works. In problems where decimal answer choices are rounded, none of the choices may work satisfactorily. Be careful not to overuse this technique.

Three calculator codes are used in the Diagnostic Test and sample tests in this book. Questions that are best approached without a calculator are labeled inactive (i). These are typically problems that contain only variables or where the computations are very simple. Questions labeled active (a) require the use of a calculator, for example to get decimal approximations for trigonometric ratios or radical expressions. Questions for which a graphing calculator is appropriate are labeled g. Many of these questions can be solved without a graphing calculator, and alternate solutions are given in these cases.

The College Board has established rules governing the use of calculators on the Mathematics Subject Tests:

- You may bring extra batteries or a backup calculator to the test. If you wish, you may bring both scientific and graphing calculators.
- Test centers are not expected to provide calculators, and test takers may not share calculators.
- Notify the test supervisor to have your score cancelled if your calculator malfunctions during the test and you do not have a backup.

TIP

Leave your cell phone at home, in our locker, or in your car!

- Certain types of devices that have computational power are **not permitted**: cell phones, pocket organizers, powerbooks and portable handheld computers, and electronic writing pads. Calculators that require an electrical outlet, make noise or "talk," or use paper tapes are also prohibited.
- You do not have to clear a graphing calculator memory before or after taking the test. However, any attempt to take notes in your calculator about a test and remove it from the room will be grounds for dismissal and cancellation of scores.

HOW THE TEST IS SCORED

There are 50 questions on the Math Level 2 Subject Test. Your raw score is the number of correct answers minus one-fourth of the number of incorrect answers, rounded to the nearest whole number. For example, if you get 30 correct answers, 15 incorrect answers, and leave 5 blank, your raw score would be $30 - \dfrac{1}{4}(15) \approx 26$, rounded to the nearest whole number.

Raw scores are transformed into scaled scores between 200 and 800. The formula for this transformation changes slightly from year to year to reflect varying test difficulty. In recent years, a raw score of 44 was high enough to transform to a scaled score of 800. Each point less in the raw score resulted in approximately 10 points less in the scaled score. For a raw score of 44 or more, the approximate scaled score is 800. For raw scores less than 44, the following formula can be used to get an approximate scaled score on the Diagnostic Test and each model test:

$S = 800 - 10(44 - R)$, where S is the approximate scaled sore and R is the rounded raw score less than 44.

The self-evaluation page for the Diagnostic Test and each model test includes spaces for you to calculate your raw score and scaled score.

STRATEGIES TO MAXIMIZE YOUR SCORE

- **Budget your time.** Although most testing centers have wall clocks, you would be wise to have a watch on your desk. Since there are 50 items on a one-hour test, you have a little over a minute per item. Typically, test items are easier near the beginning of a test, and they get progressively more difficult. Don't linger over difficult questions. Work the problems you are confident of first, and then return later to the ones that are difficult for you.
- **Guess intelligently.** As noted above, you are likely to get a higher score if you can confidently eliminate two or more answer choices, and a lower score if you can't eliminate any.
- **Read the questions carefully.** Answer the question asked, not the one you may have expected. For example, you may have to solve an equation to answer the question, but the solution itself may not be the answer.
- **Mark answers clearly and accurately.** Since you may skip questions that are difficult, be sure to mark the correct number on your answer sheet. If you change an answer, erase cleanly and leave no stray marks. Mark only one answer; an item will be graded as incorrect if more than one answer choice is marked.
- **Change an answer only if you have a good reason for doing so.** It is usually not a good idea to change an answer on the basis of a hunch or whim.
- **As you read a problem, think about possible computational shortcuts to obtain the correct answer choice.** Even though calculators simplify the computational process, you may save time by identifying a pattern that leads to a shortcut.
- **Substitute numbers to determine the nature of a relationship.** If a problem contains only variable quantities, it is sometimes helpful to substitute numbers to understand the relationships implied in the problem.

- **Think carefully about whether to use a calculator.** The College Board's guideline is that a calculator is useful or necessary in about 60% of the problems on the Level 2 Test. An appropriate percentage for you may differ significantly from this, depending on your experience with calculators, especially graphing calculators. Even if you learned the material in a highly calculator-active environment, you may discover that a problem can be done more efficiently without a calculator than with one.
- **Check the answer choices.** If the answer choices are in decimal form, the problem is likely to require the use of a calculator.

STUDY PLANS

Your first step is to take the Diagnostic Test. This should be taken under test conditions: timed, quiet, without interruption. Correct the test and identify areas of weakness using the cross-references to the Part 2 review. Use the review to strengthen your understanding of the concepts involved. Check the Part 3 graphing calculator applications to see when this approach might be optimal for you.

Ideally, you would start preparing for the test two to three months in advance. Each week, you would be able to take one sample test, following the same procedure as for the Diagnostic Test. Depending on how well you do, it might take you anywhere between 15 minutes and an hour to complete the work after you take the test. Obviously, if you have less time to prepare, you would have to intensify your efforts to complete the six sample tests, or do fewer of them.

The best way to use Parts 2 and 3 of this book is as reference material. You should look through this material quickly before you take the sample tests, just to get an idea of the range of topics covered and the level of detail. However, these parts of the book are more effectively used after you've taken and corrected a sample test.

PART 1

DIAGNOSTIC TEST

Answer Sheet
DIAGNOSTIC TEST

1 Ⓐ Ⓑ Ⓒ Ⓓ   14 Ⓐ Ⓑ Ⓒ Ⓓ   27 Ⓐ Ⓑ Ⓒ Ⓓ   40 Ⓐ Ⓑ Ⓒ Ⓓ
2 Ⓐ Ⓑ Ⓒ Ⓓ   15 Ⓐ Ⓑ Ⓒ Ⓓ   28 Ⓐ Ⓑ Ⓒ Ⓓ   41 Ⓐ Ⓑ Ⓒ Ⓓ
3 Ⓐ Ⓑ Ⓒ Ⓓ   16 Ⓐ Ⓑ Ⓒ Ⓓ   29 Ⓐ Ⓑ Ⓒ Ⓓ   42 Ⓐ Ⓑ Ⓒ Ⓓ
4 Ⓐ Ⓑ Ⓒ Ⓓ   17 Ⓐ Ⓑ Ⓒ Ⓓ   30 Ⓐ Ⓑ Ⓒ Ⓓ   43 Ⓐ Ⓑ Ⓒ Ⓓ
5 Ⓐ Ⓑ Ⓒ Ⓓ   18 Ⓐ Ⓑ Ⓒ Ⓓ   31 Ⓐ Ⓑ Ⓒ Ⓓ   44 Ⓐ Ⓑ Ⓒ Ⓓ
6 Ⓐ Ⓑ Ⓒ Ⓓ   19 Ⓐ Ⓑ Ⓒ Ⓓ   32 Ⓐ Ⓑ Ⓒ Ⓓ   45 Ⓐ Ⓑ Ⓒ Ⓓ
7 Ⓐ Ⓑ Ⓒ Ⓓ   20 Ⓐ Ⓑ Ⓒ Ⓓ   33 Ⓐ Ⓑ Ⓒ Ⓓ   46 Ⓐ Ⓑ Ⓒ Ⓓ
8 Ⓐ Ⓑ Ⓒ Ⓓ   21 Ⓐ Ⓑ Ⓒ Ⓓ   34 Ⓐ Ⓑ Ⓒ Ⓓ   47 Ⓐ Ⓑ Ⓒ Ⓓ
9 Ⓐ Ⓑ Ⓒ Ⓓ   22 Ⓐ Ⓑ Ⓒ Ⓓ   35 Ⓐ Ⓑ Ⓒ Ⓓ   48 Ⓐ Ⓑ Ⓒ Ⓓ
10 Ⓐ Ⓑ Ⓒ Ⓓ   23 Ⓐ Ⓑ Ⓒ Ⓓ   36 Ⓐ Ⓑ Ⓒ Ⓓ   49 Ⓐ Ⓑ Ⓒ Ⓓ
11 Ⓐ Ⓑ Ⓒ Ⓓ   24 Ⓐ Ⓑ Ⓒ Ⓓ   37 Ⓐ Ⓑ Ⓒ Ⓓ   50 Ⓐ Ⓑ Ⓒ Ⓓ
12 Ⓐ Ⓑ Ⓒ Ⓓ   25 Ⓐ Ⓑ Ⓒ Ⓓ   38 Ⓐ Ⓑ Ⓒ Ⓓ
13 Ⓐ Ⓑ Ⓒ Ⓓ   26 Ⓐ Ⓑ Ⓒ Ⓓ   39 Ⓐ Ⓑ Ⓒ Ⓓ

Diagnostic Test

T he diagnostic test is designed to help you pinpoint the weak spots in your background. The answer explanations that follow the test are keyed to sections of the book.

To make the best use of this diagnostic test, set aside between 1 and 2 hours so you will be able to do the whole test at one sitting. Tear out the preceding answer sheet and indicate your answers in the appropriate spaces. Do the problems as if this were a regular testing session. Review the suggestions on pages 4–5.

When finished, check your answers with those at the end of the test. For those that you got wrong, note the sections containing the material that you must review. If you do not fully understand how you arrived at some of the correct answers, you should review those sections also.

Finally, fill out the self-evaluation sheet on page 31 in order to pinpoint the topics that gave you the most difficulty.

50 questions: 1 hour

Directions: Decide which answer choice is best. If the exact numerical value is not one of the answer choices, select the closest approximation. Fill in the oval on the answer sheet that corresponds to your choice.

Notes:
(1) You will need to use a scientific or graphing calculator to answer some of the questions.
(2) You will have to decide whether to put your calculator in degree or radian mode for some problems.
(3) All figures that accompany problems are plane figures unless otherwise stated. Figures are drawn as accurately as possible to provide useful information for solving the problem, except when it is stated in a particular problem that the figure is not drawn to scale.
(4) Unless otherwise indicated, the domain of a function is the set of all real numbers for which the functional value is also a real number.

Reference Information. The following formulas are provided for your information.

Volume of a right circular cone with radius r and height h: $V = \frac{1}{3}\pi r^2 h$

Lateral area of a right circular cone if the base has circumference c and slant height is l:

$S = \frac{1}{2}cl$

Volume of a sphere of radius r: $V = \frac{4}{3}\pi r^3$

Surface area of a sphere of radius r: $S = 4\pi r^2$

Volume of a pyramid of base area B and height h: $V = \frac{1}{3}Bh$

1. A linear function, f, has a slope of -2. $f(1) = 2$ and $f(2) = q$. Find q.

    (A) 0

    (B) $\dfrac{3}{2}$

    (C) $\dfrac{5}{2}$

    (D) 3

    (E) 4

2. A function is said to be even if $f(x) = f(-x)$). Which of the following is *not* an even function?

    (A) $y = |x|$
    (B) $y = \sec x$
    (C) $y = \log x^2$
    (D) $y = x^2 + \sin x$
    (E) $y = 3x^4 - 2x^2 + 17$

3. What is the radius of a sphere, with center at the origin, that passes through point $(2,3,4)$?

    (A) 3
    (B) 3.31
    (C) 3.32
    (D) 5.38
    (E) 5.39

4. If a point (x,y) is in the second quadrant, which of the following must be true?

    I. $x < y$
    II. $x + y > 0$
    III. $\dfrac{x}{y} < 0$

    (A) only I
    (B) only II
    (C) only III
    (D) only I and II
    (E) only I and III

5. If $f(x) = x^2 - ax$, then $f(a) =$

    (A) a
    (B) $a^2 - a$
    (C) 0
    (D) 1
    (E) $a - 1$

6. The average of your first three test grades is 78. What grade must you get on your fourth and final test to make your average 80?

(A) 80
(B) 82
(C) 84
(D) 86
(E) 88

7. $\log_7 9 =$

(A) 0.89
(B) 0.95
(C) 1.13
(D) 1.21
(E) 7.61

8. If $\log_2 m = x$ and $\log_2 n = y$, then $mn =$

(A) 2^{x+y}
(B) 2^{xy}
(C) 4^{xy}
(D) 4^{x+y}
(E) cannot be determined

9. How many integers are there in the solution set of $|x - 2| \leq 5$?

(A) 0
(B) 7
(C) 9
(D) 11
(E) an infinite number

10. If $f(x) = \sqrt{x^2}$, then $f(x)$ can also be expressed as

(A) x
(B) $-x$
(C) $\pm x$
(D) $|x|$
(E) $f(x)$ cannot be determined because x is unknown.

11. The graph of $(x^2 - 1)y = x^2 - 4$ has

(A) one horizontal and one vertical asymptote
(B) two vertical but no horizontal asymptotes
(C) one horizontal and two vertical asymptotes
(D) two horizontal and two vertical asymptotes
(E) neither a horizontal nor a vertical asymptote

USE THIS SPACE FOR SCRATCH WORK

$$\frac{78 + 78 + 78 + x}{4} = 80$$
$$x = 86$$

$\dfrac{\log 9}{\log 7}$

$\log_2 m = x \qquad n = 2^y$
$m = 2^x$
$2^x \cdot 2^y = 2^{x+y}$

$x - 2 \leq 5 \qquad x - 2 \geq -5$
$x \leq 7 \quad$ or $\quad x \geq -3$

$y = \dfrac{x^2 - 4}{x^2 - 1}$

Diagnostic Test

12. $\lim\limits_{x \to \infty} \left(\dfrac{3x^2 + 4x - 5}{6x^2 + 3x + 1} \right) =$

   (A) -5

   (B) $\dfrac{1}{5}$

   (C) $\dfrac{1}{2}$

   (D) 1

   (E) This expression is undefined

$\dfrac{3x^2}{6x^2} = \dfrac{1}{2}$

13. A linear function has an x-intercept of $\sqrt{3}$ and a y-intercept of $\sqrt{5}$. The graph of the function has a slope of

   (A) -1.29

   (B) -0.77

   (C) 0.77

   (D) 1.29

   (E) 2.24

$\sqrt{3}, 0$

$0, \sqrt{5}$

$m = \dfrac{\sqrt{5} - 0}{0 - \sqrt{3}} = \dfrac{\sqrt{5}}{-\sqrt{3}}$

14. If $f(x) = 2x - 1$, find the value of x that makes $f(f(x)) = 9$.

   (A) 2

   (B) 3

   (C) 4

   (D) 5

   (E) 6

$f(f(x)) = 2(2x - 1) - 1$

$9 = 4x - 2 - 1$

$12 = 4x$

15. The plane $2x + 3y - 4z = 5$ intersects the x-axis at $(a, 0, 0)$, the y-axis at $(0, b, 0)$, and the z-axis at $(0, 0, c)$. The value of $a + b + c$ is

   (A) 1

   (B) $\dfrac{35}{12}$

   (C) 5

   (D) $\dfrac{65}{12}$

   (E) 9

16. Given the set of data 1, 1, 2, 2, 2, 3, 3, 4, which one of the following statements is true?

   (A) mean ≤ median ≤ mode

   (B) median ≤ mean ≤ mode

   (C) median ≤ mode ≤ mean

   (D) mode ≤ mean ≤ median

   (E) The relationship cannot be determined because the median cannot be calculated.

mean = 2.25

median = 2

mode = 2

17. If $\dfrac{x-3y}{x} = 7$, what is the value of $\dfrac{x}{y}$?

    (A) $-\dfrac{8}{3}$

    (B) -2

    (C) $-\dfrac{1}{2}$

    (D) $\dfrac{3}{8}$

    (E) 2

18. Find all values of x that make $\begin{vmatrix} 2 & -1 & 4 \\ 3 & 0 & 5 \\ 4 & 1 & 6 \end{vmatrix} = \begin{vmatrix} x & 4 \\ 5 & x \end{vmatrix}$.

    (A) 0
    (B) ± 1.43
    (C) ± 3
    (D) ± 4.47
    (E) 5.34

19. Suppose $f(x) = \dfrac{1}{2}x^2 - 8$ for $-4 \le x \le 4$, then the maximum value of the graph of $|f(x)|$ is

    (A) -8
    (B) 0
    (C) 2
    (D) 4
    (E) 8

20. If $\tan \theta = \dfrac{2}{3}$, then $\sin \theta =$

    (A) ± 0.55
    (B) ± 0.4
    (C) 0.55
    (D) 0.83
    (E) 0.89

21. If a and b are the domain of a function and $f(b) < f(a)$, which of the following must be true?

    (A) $b < a$
    (B) $b < a$
    (C) $a = b$
    (D) $a \ne b$
    (E) $a = 0$ or $b = 0$

22. Which of the following is perpendicular to the line $y = -3x + 7$?

    (A) $y = \dfrac{1}{-3x + 7}$

    (B) $y = 7x - 3$

    (C) $y = \dfrac{1}{3}x + 5$

    (D) $y = -\dfrac{1}{3}x + 7$

    (E) $y = 3x - 7$

23. If $f(x) = j$, where j is an integer such that $j \le x < j + 1$, and $g(x) = f(x) - |f(x)|$, what is the maximum value of $g(x)$?

    (A) -1
    (B) 0
    (C) 1
    (D) 2
    (E) j

24. If $f(x) = \dfrac{1}{\sec x}$, then

    (A) $f(x) = f(-x)$

    $f(x) = \cos(x)$

    (B) $f(\dfrac{1}{x}) = -f(x)$

    (C) $f(-x) = -f(x)$

    (D) $f(x) = f(\dfrac{1}{x})$

    (E) $f(x) = \dfrac{1}{f(x)}$

25. The polar coordinates of a point P are $(2, 240°)$. The Cartesian (rectangular) coordinates of P are

    (A) $\left(-1, -\sqrt{3}\right)$
    (B) $\left(-1, \sqrt{3}\right)$
    (C) $\left(-\sqrt{3}, -1\right)$
    (D) $\left(-\sqrt{3}, 1\right)$
    (E) $\left(1, -\sqrt{3}\right)$

    $x = r\cos\theta \quad y = r\sin\theta$
    $x = -1 \quad\quad y = \sqrt{3}$

26. The height of a cone is equal to the radius of its base. The radius of a sphere is equal to the radius of the base of the cone. The ratio of the volume of the *cone* to the volume of the *sphere* is

    (A) $\dfrac{1}{12}$

    (B) $\dfrac{1}{4}$

    (C) $\dfrac{1}{3}$

    (D) $\dfrac{1}{1}$

    (E) $\dfrac{4}{3}$

27. In how many distinguishable ways can the seven letters in the word MINIMUM be arranged, if all the letters are used each time?

    (A) 7
    (B) 42
    (C) 420
    (D) 840
    (E) 5040

28. Which of the following lines are asymptotes of the graph of $y = \dfrac{x}{x+1}$?

    I. $x = 1$
    II. $x = -1$
    III. $y = 1$

    (A) I only
    (B) II only
    (C) III only
    (D) I and II
    (E) II and III

29. What is the probability of getting at least three heads when flipping four coins?

    (A) $\dfrac{3}{16}$

    (B) $\dfrac{1}{4}$

    (C) $\dfrac{5}{16}$

    (D) $\dfrac{7}{16}$

    (E) $\dfrac{3}{4}$

USE THIS SPACE FOR SCRATCH WORK

30. The positive zero of $y = 3x^2 - 4x - 5$ is, to the nearest tenth, equal to

(A) 0.8
(B) 0.7 + 1.1*i*
(C) 0.7
(D) 2.1
(E) 2.2

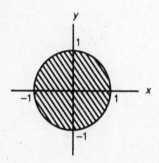

31. In the figure above, *S* is the set of all points in the shaded region. Which of the following represents the set consisting of all points $(2x, y)$, where (x, y) is a point in *S*?

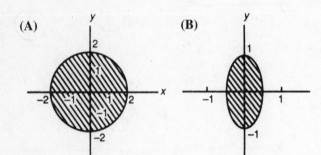

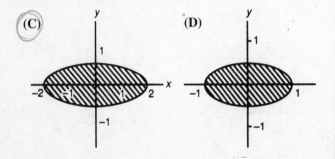

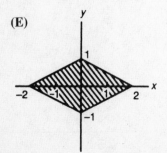

32. If a square prism is inscribed in a right circular cylinder of radius 3 and height 6, the volume inside the cylinder but outside the prism is

(A) 2.14
(B) 3.14
(C) 61.6
(D) 115.6
(E) 169.6

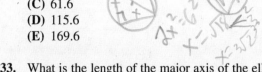

USE THIS SPACE FOR SCRATCH WORK

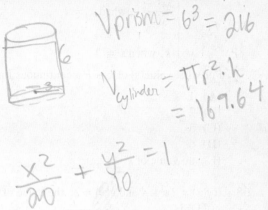

33. What is the length of the major axis of the ellipse whose equation is $10x^2 + 20y^2 = 200$?

(A) 3.16
(B) 4.47
(C) 6.32
(D) 8.94
(E) 14.14

34. The fifth term of an arithmetic sequence is 26, and the eighth term is 41. What is the first term?

(A) 3
(B) 4
(C) 5
(D) 6
(E) 7

35. $\displaystyle\sum_{j=1}^{5} 2\left(\frac{3}{2}\right)^{j-1} =$

(A) 26.125
(B) 26.375
(C) 26.500
(D) 26.625
(E) 26.875

36. If $f(x) = \dfrac{k}{x}$ for all nonzero real numbers, for what value of k does $f(f(x)) = x$?

(A) only 1
(B) only 0
(C) all real numbers
(D) all real numbers except 0
(E) no real numbers

37. $F(x) = \begin{cases} \dfrac{3x^2 - 3}{x-1}, & \text{when } x \neq 1 \\[2mm] k, & \text{when } x = 1 \end{cases}$

For what value(s) of k is F a continuous function?

(A) 1
(B) 2
(C) 3
(D) 6
(E) no value of k

38.   If $f(x) = 2x^2 - 4$ and $g(x) = 2^x$, the value of $g(f(1))$ is

(A) -4
(B) 0
(C) $\dfrac{1}{4}$
(D) 1
(E) 4

39.   If $f(x) = 3\sqrt{5x}$, what is the value of $f^{-1}(15)$?

(A) 0.65
(B) 0.90
(C) 5.00
(D) 7.5
(E) 25.98

$f(f^{-1}x) = x$

$f(15) = x$

$3\sqrt{5 \cdot 15} = x$

$f^{-1}x =$

$x = 3\sqrt{5y}$

$\left(\dfrac{x}{3}\right)^2 = 5y$

$\dfrac{\left(\dfrac{x}{3}\right)^2}{5} = y$

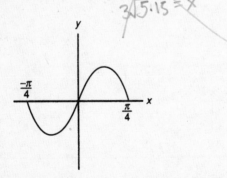

40.   Which of the following could be the equation of one cycle of the graph in the figure above?

  I. $y = \sin 4x$
  II. $y = \cos\left(4x - \dfrac{\pi}{2}\right)$
  III. $y = -\sin(4x + \pi)$

(A) only I
(B) only I and II
(C) only II and III
(D) only II
(E) I, II, and III

$\text{period} = \dfrac{2\pi}{b}$

$\dfrac{\pi}{2} = \dfrac{2\pi}{4}$

41. If $2 \sin^2 x - 3 = 3 \cos x$ and $90° < x < 270°$, the number of values that satisfy the equation is

(A) 0
(B) 1
(C) 2
(D) 3
(E) 4

42. If $A = \tan^{-1}\left(-\dfrac{3}{4}\right)$ and $A + b = 315°$, then $B =$

(A) 278.13°
(B) 351.87°
(C) −8.13°
(D) 171.87°
(E) 233.13°

43. Observers in two locations sight a rocket at a height of 12 kilometers. The angles of elevation from these locations are 80.5° and 68.0°. What is the distance between the two points?

(A) 0.85 km
(B) 4.27 km
(C) 5.71 km
(D) 20.92 km
(E) 84.50 km

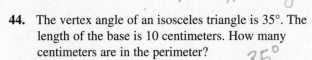

44. The vertex angle of an isosceles triangle is 35°. The length of the base is 10 centimeters. How many centimeters are in the perimeter?

(A) 16.6
(B) 17.4
(C) 20.2
(D) 43.3
(E) 44.9

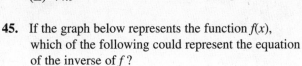

45. If the graph below represents the function $f(x)$, which of the following could represent the equation of the inverse of f?

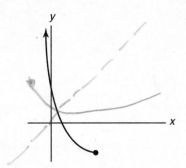

(A) $x = y^2 - 8y - 1$
(B) $x = y^2 + 11$
(C) $x = (y - 4)^2 - 3$
(D) $x = (y + 4)^2 - 3$
(E) $x = (y + 4)^2 + 3$

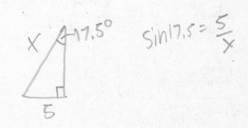

46. If $k > 4$ is a constant, how would you translate the graph of $y = x^2$ to get the graph of $y = x^2 + 4x + k$?

    (A) Left 2 units and up k units
    (B) Right 2 units and up $(k - 4)$ units
    (C) Left 2 units and up $(k - 4)$ units
    (D) Right 2 units and down $(k - 4)$ units
    (E) Left 2 units and down $(k - 4)$ units

47. If $f(x) = \log_b x$ and $f(2) = 0.231$, the value of b is

    (A) 0.3
    (B) 1.3
    (C) 13.2
    (D) 20.1
    (E) 32.5

48. If $f_{n+1} = f_{n-1} + 2f_n$ for $n = 2, 3, 4, \ldots$, and $f_1 = 1$ and $f_2 = 1$, then $f_5 =$

    (A) 7
    (B) 41
    (C) 11
    (D) 21
    (E) 17

49. Suppose $\cos \theta = u$ in $0 < \theta < \dfrac{\pi}{2}$. Then $\tan \theta =$

    (A) 1

    (B) $\dfrac{1}{\sqrt{1 - u^2}}$

    (C) $\dfrac{u}{\sqrt{1 - u^2}}$

    (D) $\sqrt{1 - u^2}$

    (E) $\dfrac{\sqrt{1 - u^2}}{u}$

50. If $[x]$ is defined to represent the greatest integer less than or equal to x, and $f(x) = \left| x - [x] - \dfrac{1}{2} \right|$, what is the period of $f(x)$?

    (A) $\dfrac{1}{2}$
    (B) 1
    (C) 2
    (D) 4
    (E) f is not a periodic function.

USE THIS SPACE FOR SCRATCH WORK

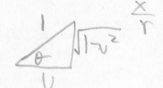

STOP

If there is still time remaining, you may review your answers.

Answer Key

DIAGNOSTIC TEST

1. **A**	14. **B**	27. **C**	40. **E**
2. **D**	15. **B**	28. **E**	41. **D**
3. **E**	16. **C**	29. **C**	42. **B**
4. **E**	17. **C**	30. **D**	43. **C**
5. **C**	18. **D**	31. **C**	44. **D**
6. **D**	19. **E**	32. **C**	45. **C**
7. **C**	20. **A**	33. **D**	46. **C**
8. **A**	21. **D**	34. **D**	47. **D**
9. **D**	22. **C**	35. **B**	48. **E**
10. **D**	23. **B**	36. **D**	49. **E**
11. **C**	24. **A**	37. **D**	50. **B**
12. **C**	25. **A**	38. **C**	
13. **A**	26. **B**	39. **C**	

ANSWERS EXPLAINED

The following explanations are keyed to the review portions of this book. The number in brackets after each explanation indicates the appropriate section in the Review of Major Topics (Part 2). If a problem can be solved using algebraic techniques alone, [algebra] appears after the explanation, and no reference is given for that problem in the Self-Evaluation Chart at the end of the test.

> In these solutions the following notation is used:
>
> i: calculator unnecessary
> a: calculator helpful or necessary
> g: graphing calculator helpful or necessary

1. i **(A)** $f(1) = 2$ means that the line goes through point $(1,2)$. $f(2) = q$ means that the line goes through point $(2,q)$. Slope $= \dfrac{\Delta y}{\Delta x} = \dfrac{q-2}{2-1} = -2$ implies $-2 = \dfrac{q-2}{1}$, so $q = 0$.
   [1.2]

2. g **(D)** Even functions are symmetric about the y-axis. Graph each answer choice to see that Choice D is not symmetric about the y-axis.

   An alternative solution is to use the fact that $\sin x \neq \sin(-x)$, from which you deduce the correct answer choice. [1.1]

> **TIP**
>
> Properties of even and odd functions:
> Even + even is always an even function.
> Odd + odd is always an odd function.
> Odd × even is always an odd function.

3. a **(E)** Since the radius of a sphere is the distance between the center, $(0,0,0)$, and a point on the surface, $(2,3,4)$, use the distance formula in three dimensions to get

   $$\sqrt{(2-0)^2 + (3-0)^2 + (4-0)^2} = \sqrt{29}$$

   Use your calculator to find $\sqrt{29} \approx 5.39$. [2.2]

4. i **(E)** A point in the second quadrant has a negative x-coordinate and a positive y-coordinate. Therefore, $x < y$, and $\dfrac{x}{y} < 0$ must be true, but $x + y$ can be less than or equal to zero. The correct answer is E. [1.1]

5. i **(C)** $f(a)$ means to replace x in the formula with an a. Therefore, $f(a) = a^2 - a \cdot a = 0$. [1.1]

6. a **(D)** Since the average of your first three test grades is 78, each test grade could have been a 78. If x represents your final test grade, the average of the four test grades is $\dfrac{78+78+78+x}{4}$, which is to be equal to 80. Therefore, $\dfrac{234+x}{4} = 80$.

   $234 + x = 320$. So $x = 86$. [4.1]

7. a **(C)** Use the change-of-base theorem and your calculator to get:

$$\log_7 9 = \frac{\log_{10} 9}{\log_{10} 7} \approx \frac{0.9542}{0.8451} \approx 1.13. \ [1.4]$$

8. i **(A)** Add the two equations: $\log_2 m + \log_2 n = x + y$, which becomes $\log_2 mn = x + y$ (basic property of logs). $2^{x+y} = mn$. [1.4]

9. g **(D)** Plot the graph of $y = abs(x - 2) - 5$ in the standard window that includes both x–intercepts. You can count 11 integers between –3 and 7 if you include both endpoints.

The inequality $|x - 2| \le 5$ means that x is less than or equal to 5 units away from 2. Therefore, $-3 \le x \le 7$, and there are 11 integers in this interval. [1.6]

10. i **(D)** $\sqrt{x^2}$ indicates the need for the *positive* square root of x^2. Therefore, $\sqrt{x^2} = x$ if $x \ge 0$ and $\sqrt{x^2} = -x$ if $x < 0$. This is just the definition of absolute value, and so $\sqrt{x^2} = |x|$ is the only answer for all values of x. [1.6]

11. g **(C)** Solve for y, and plot the graph of $y = \dfrac{x^2 - 4}{x^2 - 1}$ in the standard window to observe two vertical and one horizontal asymptotes.

An alternative solution is to use the facts that $y = \dfrac{x^2 - 4}{x^2 - 1}$ has vertical asymptotes when the denominator is zero, i.e., when $x = \pm 1$, and a horizontal asymptote of $y = 1$ as $x \to \infty$. [1.5]

12. g **(C)** Enter the given expression into Y_t and key in TBLSET with TblStart = 0 and ΔTbl = 10. Observe Y_t approach 0.5 as x gets larger.

Divide the numerator and denominator of the expression by x^2 and observe that the expression approaches $\dfrac{1}{2}$ as $x \to \infty$. [1.5]

13. a **(A)** $y = mx + b$. Use the x-intercept to get $0 = \sqrt{3}m + b$, and the y-intercept to get

$\sqrt{5} = 0 \cdot m + b$. Therefore, $0 = \sqrt{3}m + \sqrt{5}$ and $m = -\dfrac{\sqrt{5}}{\sqrt{3}} \approx -1.29$. [1.2]

14. i **(B)** $f(f(x)) = 2(2x - 1) - 1 = 4x - 3$. Solve $4x - 3 = 9$ to get $x = 3$. [1.1]

15. i **(B)** Substituting the points into the equation gives $a = \dfrac{5}{2}, b = \dfrac{5}{3}$, and $c = -\dfrac{5}{4}$. [2.2]

16. i **(C)** Mode = 2, median = $\dfrac{2+2}{2} = 2$, mean = $\dfrac{2+6+6+4}{8} = \dfrac{18}{8} = 2.25$.

Thus, median $\le$ mode $\le$ mean. [4.1]

17. i **(C)** Multiply $\dfrac{x-3y}{x} = 7$ through by x to get $x - 3y = 7x$. Subtract x from both sides

to get $-3y = 6x$. Divide through by $6y$ so that $\dfrac{x}{y}$ will be on one side of the equals sign.

This gives $\dfrac{x}{y} = -\dfrac{3}{6} = -\dfrac{1}{2}$. [algebra]

18. g **(D)** Enter the 3 by 3 matrix into the graphing calculator and evaluate its determinant as 0. The 2 by 2 matrix on the right side of the equation has the determinant $x^2 - 20$. Solve this for x to get ± 4.47. [3.3]

19. g **(E)** Plot the graph of $y = abs((1/2)x^2 - 8)$ in a $[-4,4]$ by $[-10,10]$ window and observe that the maximum value is -8 (at $x = 0$).

An alternative solution is to recognize that the graph of f is a parabola that is symmetric about the y – axis and opens up. The minimum value $y = -8$ occurs when $x = 0$, so 8 is the maximum value of $|f(x)|$. [1.2]

20. a **(A)** Press $\sin(\tan^{-1}(2/3)) \approx 0.55$. Tan is positive in the first and third quadrants, so θ is a first or third quadrant angle. The sine of a first quadrant angle is positive, but the sine of a third quadrant angle is negative. The correct answer choice is therefore ± 0.55. [1.3]

21. i **(D)** If $a = b$, then $f(a) = f(b)$. Since $f(a) \neq f(b)$, it follows that $a \neq b$. [1.1]

22. i **(C)** The slope of the given line is -3. Therefore, the slope of a perpendicular line is the negative reciprocal, or $\dfrac{1}{3}$. [1.2]

23. g **(B)** The function f is the greatest integer function. Plot the graph of $y = int(x) - abs(int(x))$ in the standard window and observe that the maximum value of y is 0.

An alternative solution is to observe that $f(x) = |f(x)|$ if $x \geq 0$, and $f(x) = -|f(x)|$ if $x < 0$. In the latter case, $g(x) = 2f(x) < 0$, so the maximum is 0. [1.6]

24. g **(A)** Plot the graph of $y = \dfrac{1}{\cos x}$ and observe that it is symmetric about the y-axis.

Hence $f(x) = \sec x$ is an even function and $f(x) = f(-x)$. An alternative solution is to recall that $\cos x$ is even, so its reciprocal is also even. [1.1, 1.3]

25. i **(A)** From the figure $\left(-1, -\sqrt{3}\right)$. [2.1]

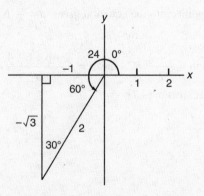

26. i **(B)** Since $r = h$ in the cone,

$$\frac{\text{Volume of cone}}{\text{Volume of sphere}} = \frac{\frac{1}{3}\pi r^2 h}{\frac{4}{3}\pi r^3} = \frac{\frac{1}{3}\pi r^3}{\frac{4}{3}\pi r^3} = \frac{1}{4}. \quad [2.2]$$

27. a **(C)** The word MINIMUM contains 7 letters, which can be permuted 7! ways. The 3 M's can be permuted 3! ways, and the 2 I's can be permuted in 2! ways, so only $\frac{1}{3!2!} = \frac{1}{12}$ permutations look different from each other. Therefore, there are $\frac{7!}{12} = 420$ distinguishable ways the letters can be arranged. [3.1]

28. i **(E)** Vertical asymptotes occur when the denominator is zero, so $x = -1$ is the only vertical asymptote. Since $\lim\limits_{x \to \pm\infty} \frac{x}{x+1} = 1$ the y asymptote is $y = 1$. [1.5]

29. a **(C)** Since each of the 4 flips has 2 possible outcomes (heads or tails), these are $2^4 = 16$ outcomes in the sample space. At least 3 heads means 3 or 4 heads. $\binom{4}{3} = 4$ ways to get 3 heads. $\binom{4}{4} = 1$ way to get 4 heads and $\frac{4+1}{16} = \frac{5}{16}$. [3.1]

30. g **(D)** Use the quadratic formula program on your graphing calculator to get both zeros and choose the positive one. [1.2]

31. i **(C)** Since the y values remain the same but the x values are doubled, the circle is stretched along the x-axis. [2.1]

32. a **(C)** Volume of cylinder $= \pi r^2 h = \pi \cdot 9 \cdot 6 = 54\pi$. Volume of square prism $= Bh$, where B is the area of the square base, which is $3\sqrt{2}$ on a side. Thus, $Bh = (3\sqrt{2})^2 \cdot 6 = 108$. Therefore, the desired volume is $54\pi - 108$, which (using your calculator) is approximately 61.6. [2.2]

33. a **(D)** Divide both sides of the equation by 200 to write the equation in standard form $\frac{x^2}{20} + \frac{y^2}{10} = 1$. The length of the major axis is $2\sqrt{20} \approx 8.94$. [2.1]

34. i **(D)** There are three constant differences between the fifth and eighth terms. Since $41 - 26 = 15$, the constant difference is 5. The fifth term, 26, is four constant differences (20) more than the first term. Therefore, the first term is $26 - 20 = 6$. [3.4]

35. a **(B)** Use the LIST/OPS/seq command to generate the sequence $2\left(\frac{3}{2}\right)^{j-2}$ as follows:

seq(2(3/2)^(x – 1), x,1,5). Then use LIST/MATH/sum (Ans) to get the correct answer choice. This result can be achieved with the single command LIST/MATH/sum(LIST/OPS/seq(2(3/2)∧(x – 1), x,1,5)).

An alternative solution is to use your calculator to evaluate each term and add them. [3.4]

36. i **(D)** $f(f(x)) = \dfrac{k}{k/x} = k \cdot \dfrac{x}{k} = x$. Since k is in the denominator, it cannot equal 0. [1.1]

37. g **(D)** Enter $(3x^2 - 3)/(x - 1)$ into Y_1 and key TBLSET. Set Indpnt to Ask and key TABLE. Enter values of x progressively closer to 1 (e.a. .9, .99, .999, etc.) and observe that Y_1 gets progressively closer to 6, so choose $k = 6$.

An alternative solution is to factor the numerator to $3(x + 1)(x - 1)$, divide out $x - 1$ from the numerator and denominator. As $x \to 1$, $3(x + 1) \to 6$. [1.5]

38. i **(C)** $f(1) = 2 - 4 = -2$, and $g(-2) = 2^{-2} = \dfrac{1}{4}$. [1.1]

39. i **(C)** $f^{-1}(15)$ is the value of x that makes $3\sqrt{5x}$ equal to 15. Set $3\sqrt{5x} = 15$, divide both sides by 3 to get $\sqrt{5x} = 5$. [1.1]

40. g **(E)** Plot the graphs of all three functions in a $\left[-\dfrac{\pi}{4}, \dfrac{\pi}{4} \right]$ by $[-2,2]$ window and observe that they coincide.

An alternative solution is to deduce facts about the graphs from the equations. All three equations indicate graphs that have period $\dfrac{\pi}{2}$. The graph of equation I is a normal sine curve. The graph of equation II is a cosine curve with a phase shift right of $\dfrac{\pi}{8}$, one-fourth of the period. Therefore, it fits a normal sine curve. The graph of equation III is a sine curve that has a phase shift left of $\dfrac{\pi}{4}$, one-half the period, and reflected through the x-axis. This also fits a normal sine curve. [1.3]

41. g **(D)** With your calculator in degree mode, plot the graphs of $2(\sin(x))^2 - 3$ and $y = 3\cos x$ in a $[90,270]$ by $[-3,0]$ window and observe that the graphs intersect in 3 places.

An alternative solution is to distribute and transform the equation to read: $2\cos^2 x + 3\cos x + 1 = 0$. This factors to $(2\cos x + 1)(\cos x + 1) = 0$, so $\cos x = -\dfrac{1}{2}$ or $\cos x = -1$.

For $90° < x < 270°$, there are three solutions: $x = 120°, 240°, 180°$. [1.3]

42. g **(B)** With your calculator in degree mode, evaluate $315° - \tan^{-1}(-3/4)$ to get the correct answer choice. [1.3]

43. a **(C)** The problem information is summarized in the figure below.

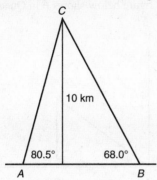

Points A and B represent the two observers. Point C is the base of the altitude from the rocket to the ground. We know that $\tan 80.5° = \dfrac{10}{AC}$ and $\tan 68.0° = \dfrac{10}{CB}$. Therefore,

$$AB = AC + CB = \frac{10}{\tan 80.5°} + \frac{10}{\tan 68.0°} \approx 5.71.\ [1.3]$$

44. a **(D)** Drop the altitude from the vertex to the base. The altitude bisects both the vertex angle and the base, cutting the triangle into two congruent right triangles.

Since $\sin 17.5° = \dfrac{5}{\text{leg}}$, leg $= \dfrac{5}{\sin 17.5°} \approx \dfrac{5}{0.3007} \approx 16.628$ cm and the perimeter $=$ 43.3 cm. [1.3]

45. i **(C)** The given graph looks like the left half of a parabola with vertex $(4,-3)$ (using the values given in the answer choices as guides) that opens up. The equation of such a parabola is $y = (x - 4)^2 - 3$. The vertex of the inverse is $(-3,4)$, so its equation is $x = (y - 4)^2 - 3$. [1.1]

46. i **(C)** Complete the square of $x^2 + 4x + k$ by adding and subtracting 4 to get the translated function of $y = (x + 2)^2 + (k - 4)$. Translate $y = x^2$ left 2 units and up $(k - 4)$ units. [2.1]

47. a **(D)** $f(2) = \log_b 2 = 0.231$. Therefore, $b^{0.231} = 2$, and so $b = 2^{1/0.231}$, which (using your calculator) is approximately 20.1. [1.4]

48. i **(E)** Let $n = 2, f_3 = 3$; then let $n = 3, f_4 = 7$; and finally let $n = 4, f_5 = 17$. [3.4]

49. i **(E)** Since $0 < \theta < \dfrac{\pi}{2}$, the figure below shows θ in Quadrant I with $\cos \theta = u$.

From the figure, $\tan \theta = \dfrac{\sqrt{1 - u^2}}{u}$. [1.3]

50. a **(B)** Plot the graph of $y = abs(x - int(x) - 1/2)$ in a $[-3,3]$ by $[-1,1]$ window. Inspection of the graph leads to the conclusion that the period is 1. [1.6]

Self-Evaluation Chart for Diagnostic Test

Subject Area	Questions and Review Section							Right	Number Wrong	Omitted
Algebra and Functions (27 questions)	1 1.2	2 1.1	4 1.1	5 1.1	7 1.4	8 1.4		5	0	0
	9 1.6	10 1.6	11 1.5	12 1.5	13 2.1	14 1.1	17	7	0	0
	19 1.2	21 1.1	22 1.2	23 1.6	24 1.1 1.3	28 1.5	30 1.2	6	1	0
	36 1.1	37 1.5	38 1.1	39 1.1	45 1.1	47 1.4	50 1.6	6	1	0
Trigonometry (8 questions)	20 1.3	24 1.3	40 1.3	41 1.3	42 1.3	43 1.3		4	1	1
	44 1.3	49 1.3						2	0	0
Coordinate and Three-Dimensional Geometry (8 questions)	③ 2.2	15 2.2	25 2.1	26 2.2	31 2.1	32 2.2		3	0	3
	33 2.1	46 2.1						2	0	0
Numbers and Operations (6 questions)	⑱ 3.3	27 3.1	㉙ 3.1	34 3.4	35 3.4	48 3.4		III	II	I
Data Analysis, Statistics, and Probability (2 questions)	6 4.1	16 4.1						1	1	0
TOTALS								39	6	5

Evaluate Your Performance Diagnostic Test	
Rating	**Number Right**
Excellent	41–50
Very good	33–40
Above average	25–32
Average	15–24
Below average	Below 15

Calculating Your Score

Raw score R = number right $- \dfrac{1}{4}$ (number wrong), rounded = _____ 38

Approximate scaled score $S = 800 - 10(44 - R)$ = _____ 740

If $R \geq 44$, $S = 800$.

PART 2

REVIEW OF MAJOR TOPICS

This part reviews the mathematical concepts and techniques for the topics covered in the Math Level 2 Subject Test. A sound understanding of these concepts certainly will improve your score, whether you use a scientific calculator or a graphing calculator. The techniques discussed may help you save time solving some of the problems without a calculator at all. For problems requiring computational power, techniques are described that will help you use your calculator in the most efficient manner. Methods for solving a variety of problems using a graphing calculator are discussed in Part 3.

Your classroom experience will guide your decisions about whether to use a graphing calculator and how best to use one. If you have been through a secondary mathematics program that attached equal importance to graphical, tabular, and algebraic presentations, then you probably will rely on your graphing calculator as your primary tool to help you find solutions. However, if you went through a more traditional mathematics program, where algebra and algebraic techniques were stressed, it may be more natural for you to use a graphing calculator only after considering other approaches.

Functions

CHAPTER 1

- Overview
- Polynomial Functions
- Trigonometric Functions and Their Inverses
- Exponential and Logarithmic Functions
- Rational Functions and Limits
- Miscellaneous Functions

1.1 Overview

DEFINITIONS

A *relation* is a set of ordered pairs. A *function* is a relation such that for each first element there is one and only one second element. The set of numbers that make up all first elements of the ordered pairs is called the *domain* of the function, and the resulting set of second elements is called the *range* of the function.

Example 1

$$\{(1,2),(3,4),(5,6),(6,1),(2,2)\}$$

This is a function because every ordered pair has a different first element. Domain = {1,2,3,5,6}. Range = {1,2,4,6}.

Example 2

$$f(x) = 3x + 2$$

This is a function because for each value substituted for x there is one and only one value for $f(x)$. Domain = {all real numbers}. Range = {all real numbers}.

Example 3

$$\{(1,2),(3,2),(1,4)\}$$

This is a relation but *not* a function because when the first element is 1, the second element can be either 2 or 4. Domain = {1,3}. Range = {2,4}.

> ### Example 4
>
> $\{(x,y):y^2 = x\}$
>
> [This should be read, "The set of all ordered pairs (x,y) such that $y^2 = x$."] This is a relation but *not* a function because, for each nonnegative number that is substituted for x, there are two values for y. For example, $x = 4$, $y = +2$, or $y = -2$. Domain = {all nonnegative real numbers}. Range = {all real numbers}.
>
> The expressions $f = \{(x,y):y = x^2\}$ and $f(x) = x^2$ both name the same function; f is the rule that pairs any number with its square. Thus, $f(x) = x^2$, $f(a) = a^2$, $f(z) = z^2$ all name the same function. The symbol $f(2)$ is the value of the function f when $x = 2$. Thus, $f(2) = 4$.

TIP

Looking for answers? All answers to exercises appear at the end of each section. Resist the urge to peek before solving the problems on your own.

EXERCISES

1. If $\{(3,2),(4,2),(3,1),(7,1),(2,3)\}$ is to be a function, which one of the following must be removed from the set?

   (A) $(3,2)$
   (B) $(4,2)$
   (C) $(2,3)$
   (D) $(7,1)$
   (E) none of the above

2. For $f(x) = 3x^2 + 4$, $g(x) = 2$, and h $\{(1,1), (2,1), (3,2)\}$,

   (A) f is the only function
   (B) h is the only function
   (C) f and g are the only functions
   (D) g and h are the only functions
   (E) f, g, and h are all functions

3. What value(s) must be excluded from the domain of $f = \left\{(x, y): y = \dfrac{x+2}{x-2}\right\}$?

   (A) -2
   (B) 0
   (C) 2
   (D) 2 and -2
   (E) no value

COMBINING FUNCTIONS

If f and g name two functions, the following rules apply:

$$(f + g)(x) = f(x) + g(x)$$

$$(f - g)(x) = f(x) - g(x)$$

$$(f \cdot g)(x) = f(x) \cdot g(x)$$

$$\left(\frac{f}{g}\right)(x) = \frac{f(x)}{g(x)}$$

if and only if $g(x) \neq 0$

$$(f \circ g)(x) = f(g(x))$$

(this is called the *composition* of functions).

Example

If $f(x) = 3x - 2$ and $g(x) = x^2 - 4$, then indicate the function that represents

(A) $(f + g)(x)$
(B) $(f - g)(x)$
(C) $(f \cdot g)(x)$
(D) $\dfrac{f}{g}(x)$
(E) $(f \circ g)(x)$
(F) $(g \circ f)(x)$

SOLUTIONS

(A) $(f + g)(x) = f(x) + g(x)$
$$= (3x - 2) + (x^2 - 4) = x^2 + 3x - 6$$

(B) $(f - g)(x) = f(x) - g(x)$
$$= (3x - 2) - (x^2 - 4) = -x^2 + 3x + 2$$

(C) $(f \cdot g)(x) = f(x) \cdot g(x)$
$$= (3x - 2)(x^2 - 4)$$
$$= 3x^3 - 2x^2 - 12x + 8$$

(D) $\dfrac{f}{g}(x) = \dfrac{f(x)}{g(x)} = \dfrac{3x - 2}{x^2 - 4}$ and $x \neq \pm 2$

(E) $(f \circ g)(x) = f(x) \circ g(x)$
$$= f(g(x)) = 3(g(x)) - 2$$
$$= 3(x^2 - 4) - 2 = 3x^2 - 14$$

(F) $(g \circ f)(x) = g(x) \circ f(x)$
$$= g(f(x)) = (f(x))^2 - 4$$
$$= (3x - 2)^2 - 4 = 9x^2 - 12x$$

(Notice that the composition of functions is not commutative.)

CAREFUL!

$(f \circ g)(x)$ and $(g \circ f)(x)$ *are not* the same!

EXERCISES

1. If $f(x) = 3x^2 - 2x + 4$, $f(-2) =$

    (A) −12
    (B) −4
    (C) −2
    (D) 12
    (E) 20

2. If $f(x) = 4x - 5$ and $g(x) = 3^x$, then $f(g(2)) =$

    (A) 3
    (B) 9
    (C) 27
    (D) 31
    (E) none of the above

3. If $f(g(x)) = 4x^2 - 8x$ and $f(x) = x^2 - 4$, then $g(x) =$

(A) $4 - x$
(B) x
(C) $2x - 2$
(D) $4x$
(E) x^2

4. What values must be excluded from the domain of $\left(\dfrac{f}{g}\right)(x)$ if $f(x) = 3x^2 - 4x + 1$ and $g(x) = 3x^2 - 3$?

(A) 0
(B) 1
(C) 3
(D) both ± 1
(E) no values

5. If $g(x) = 3x + 2$ and $g(f(x)) = x$, then $f(2) =$

(A) 0
(B) 1
(C) 2
(D) 6
(E) 8

6. If $p(x) = 4x - 6$ and $p(a) = 0$, then $a =$

(A) -6

(B) $-\dfrac{3}{2}$

(C) $\dfrac{3}{2}$

(D) $\dfrac{2}{3}$

(E) 2

7. If $f(x) = e^x$ and $g(x) = \sin x$, then the value of $(f \circ g)(\sqrt{2})$ is

(A) -0.01
(B) -0.8
(C) 0.34
(D) 1.8
(E) 2.7

INVERSES

The *inverse* of a function f, denoted by f^{-1}, is a relation that has the property that $f(x) \circ f^{-1}(x) = f^{-1}(x) \circ f(x) = x$, where f^{-1} is not necessarily a function.

Example 1

$f(x) = 3x + 2$. Is $y = \dfrac{x - 2}{3}$ the inverse of f?

To answer this question assume that $f^{-1}(x) = \dfrac{x-2}{3}$ and verify that $f(x) \circ f^{-1}(x) = x$.

To verify this, proceed as follows:

$$f(x) \circ f^{-1}(x) = f(f^{-1}(x))$$
$$= f\left(\frac{x-2}{3}\right) = 3\left(\frac{x-2}{3}\right) + 2 = x$$

and

$$f^{-1}(x) \circ f(x) = f^{-1}(f(x)) = f^{-1}(3x+2)$$
$$= \frac{(3x+2)-2}{3} = x.$$

Since $f(x) \circ f^{-1}(x) \circ f^{-1}(x) \circ f(x) = x$, $\dfrac{x-2}{3}$ is the inverse of f.

Example 2

$f = \{(1,2),(2,3),(3,2)\}$ **Find the inverse.**

$$f^{-1} = \{(2,1),(3,2),(2,3)\}$$

> **CAREFUL!**
>
> The inverse of a function need not be a function.

To verify this, check $f \circ f^{-1}$ and $f^{-1} \circ f$ term by term.

$(f \circ f^{-1})^{(x)} = f(f^{-1}(x))$; when $x = 2$, $f(f^{-1}(2)) = f(1) = 2$

when $x = 3$, $f(f^{-1}(3)) = f(2) = 3$

when $x = 2$, $f(f^{-1}(2)) = f(3) = 2$

Thus, for each x, $f(f^{-1}(x)) = x$.

$(f^{-1} \circ f)^{(x)} = f^{-1} \circ (f(x))$; when $x = 1$, $f^{-1}(f(1)) = f^{-1}(2) = 1$

when $x = 2$, $f^{-1}(f(2)) = f^{-1}(3) = 2$

when $x = 3$, $f^{-1}(f(3)) = f^{-1}(2) = 3$

Thus, for each x, $f^{-1}(f(x)) = x$. In this case f^{-1} is *not* a function.

If the point with coordinates (a,b) belongs to a function f, then the point with coordinates (b,a) belongs to the inverse of f. Because this is true of a function and its inverse, the graph of the inverse is the reflection of the graph of f about the line $y = x$.

Example 3

f^{-1} is *not* a function.

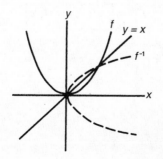

> **LOOK!**
>
> Graphs of inverses are reflections about the line $y = x$.

Example 4

f^{-1} **is a function.**

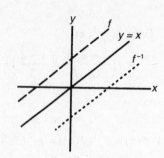

As can be seen from the above examples, if the graph of f is given, the graph of f^{-1} is the image obtained by folding the graph of f about the line $y = x$. Algebraically the equation of the inverse of f can be found by interchanging the variables.

Example 5

$f = \{(x,y):y = 3x + 2\}$. **Find** f^{-1}.

In order to find f^{-1}, interchange x and y and solve for y: $x = 3y + 2$, which becomes $y = \dfrac{x-2}{3}$.

Thus,

$$f^{-1} = \left\{ (x,y) : y = \frac{x-2}{3} \right\}$$

Example 6

$f = \{(x,y):y = x^2\}$. **Find** f^{-1}.

Interchange x and y: $x = y^2$.

Solve for $y = y = \pm\sqrt{x}$.

Thus, $f^{-1} = \left\{ (x,y) : y = \pm\sqrt{x} \right\}$, which is *not* a function.

The inverse of any function f can always be made a function by limiting the domain of f. In Example 6 the domain of f could be limited to all nonnegative numbers or all nonpositive numbers. In this way f^{-1} would become either $y = +\sqrt{x}$ or $y = -\sqrt{x}$, both of which are functions.

Example 7

$f = \{(x,y):y = x^2 \text{ and } x \geq 0\}$. **Find** f^{-1}.

$f^{-1} = \{(x,y):x = y^2 \text{ and } y \geq 0\}$, which can also be written as

$$f^{-1} = \left\{ (x,y) : y = +\sqrt{x} \right\}$$

Here f^{-1} is a function.

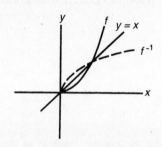

Example 8

$f = \{(x,y):y = x^2 \text{ and } x \le 0\}$. Find f^{-1}.

$f^{-1} = \{(x,y):x = y^2 \text{ and } y \le 0\}$, which can also be written as

$$f^{-1} = \left\{(x,y): y = -\sqrt{x}\right\}.$$

Here f^{-1} is a function.

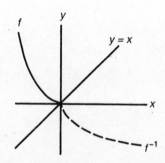

EXERCISES

1. If $f(x) = 2x - 3$, the inverse of f, f^{-1}, could be represented by

   (A) $f^{-1}(x) = 3x - 2$

   (B) $f^{-1}(x) = \dfrac{1}{2x - 3}$

   (C) $f^{-1}(x) = \dfrac{x - 2}{3}$

   (D) $f^{-1}(x) = \dfrac{x + 2}{3}$

   (E) $f^{-1}(x) = \dfrac{x + 3}{2}$

2. If $f(x) = x$, the inverse of f, f^{-1}, could be represented by

   (A) $f^{-1}(x) = x$

   (B) $f^{-1}(x) = 1$

   (C) $f^{-1}(x) = \dfrac{1}{x}$

   (D) $f^{-1}(x) = y$

   (E) f^{-1} does not exist

3. The inverse of $f = \{(1,2),(2,3),(3,4),(4,1),(5,2)\}$ would be a function if the domain of f is limited to

(A) $\{1,3,5\}$
(B) $\{1,2,3,4\}$
(C) $\{1,5\}$
(D) $\{1,2,4,5\}$
(E) $\{1,2,3,4,5\}$

4. Which of the following could represent the equation of the inverse of the graph in the figure?

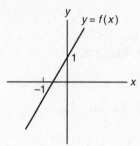

(A) $y = -2x + 1$

(B) $y = 2x + 1$

(C) $y = \dfrac{1}{2}x + 1$

(D) $y = \dfrac{1}{2}x - 1$

(E) $y = \dfrac{1}{2}x - \dfrac{1}{2}$

ODD AND EVEN FUNCTIONS

A relation is said to be *even* if $(-x,y)$ is in the relation whenever (x,y) is. If the relation is defined by an equation, it is even if $(-x,y)$ satisfies the equation whenever (x,y) does. If the relation is a function f, it is even if $f(-x) = f(x)$ for all x in the domain of f. The graph of an even relation or function is symmetric with respect to the y axis.

Example 1

$\{(1,0),(-1,0),(3,0),(-3,0),(5,4),(-5,4)\}$ is an even relation because $(-x,y)$ is in the relation whenever (x,y) is.

Example 2

$x^4 + y^2 = 10$ is an even relation because $(-x)^4 + y^2 = x^4 + y^2 = 10$.

Example 3

$f(x) = x^2$ and $f(-x) = (-x)^2 = x^2$.

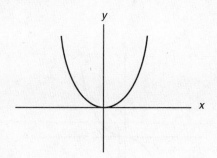

Example 4

$f(x) = |x|$ and $f(-x) = |-x| = |-1 \cdot x| = |-1| \cdot |x| = |x|$.

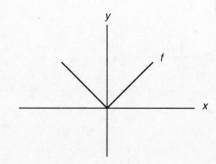

A relation is said to be *odd* if $(-x,-y)$ is in the relation whenever (x,y) is. If the relation is defined by an equation, it is odd if $(-x,-y)$ satisfies the equation whenever (x,y) does. If the relation is a function f, it is odd if $f(-x) = -f(x)$ for all x in the domain of x. The graph of an odd relation or function is symmetric with respect to the origin.

> **TIP**
>
> An odd relation is reflected through the origin.

Example 5

$\{(5,3),(-5,-3),(2,1),(-2,-1),(-10,8), (10,-8)\}$ is an odd relation because $(-x,-y)$ is in the relation whenever (x,y) is.

Example 6

$x^4 + y^2 = 10$ is an odd relation because $(-x)^4 + (-y)^2 = x^4 + y^2 = 10$. Note that $x^4 + y^2 = 10$ is both even and odd.

Example 7

$f(x) = x^3$ **and** $f(-x) = (-x)^3 = -x^3$.

Therefore, $f(-x) = x^3 = -f(x)$.

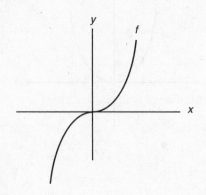

Example 8

$f(x) = \dfrac{1}{x}$ **and** $f(-x) = \dfrac{1}{-x}$.

Therefore, $f(-x) = \dfrac{1}{x} = -f(x)$.

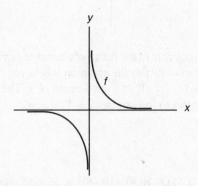

The sum of even functions is even. The sum of odd functions is odd. The product of an even function and an odd function is odd.

EXERCISES

1. Which of the following relations are *even*?

   I. $y = 2$
  II. $f(x) = x$
 III. $x^2 + y^2 = 1$

(A) only I
(B) only I and II
(C) only II and III
(D) only I and III
(E) I, II, and III

2. Which of the following relations are *odd*?

    I. $y = 2$
    II. $y = x$
    III. $x^2 + y^2 = 1$

    (A) only II
    (B) only I and II
    (C) only I and III
    (D) only II and III
    (E) I, II, and III

3. Which of the following relations are both *odd* and *even*?

    I. $x^2 + y^2 = 1$
    II. $x^2 - y^2 = 0$
    III. $x + y = 0$

    (A) only III
    (B) only I and II
    (C) only I and III
    (D) only II and III
    (E) I, II, and III

4. Which of the following functions is neither *odd* nor *even*?

    (A) $\{(1,2),(4,7),(-1,2),(0,4),(-4,7)\}$
    (B) $\{(1,2),(4,7),(-1,-2),(0,0),(-4,-7)\}$
    (C) $\{(x,y):y = x^3 - 1\}$
    (D) $\{(x,y):y = x^2 - 1\}$
    (E) $f(x) = -x$

Answers and Explanations

In these solutions the following notation is used:

> i: calculator unnecessary
> a: calculator helpful or necessary
> g: graphing calculator helpful or necessary

Definitions

1. i **(A)** Either (3,2) or (3,1), which is not an answer choice, must be removed so that 3 will be paired with only one number.

2. i **(E)** For each value of x there is only one value for y in each case. Therefore, f, g, and h are all functions.

3. i **(C)** Since division by zero is forbidden, x cannot equal 2.

Combining Functions

1. i **(E)** $f(-2) = 3(-2)^2 - 2(-2) + 4 = 20$.

2. i **(D)** $g(2) = 3^2 = 9$. $f(g(2)) = f(9) = 31$.

3. i **(C)** To get from $f(x)$ to $f(g(x))$, x^2 must become $4x^2$. Therefore, the answer must contain $2x$ since $(2x)^2 = 4x^2$.

4. i **(D)** $g(x)$ cannot equal 0. Therefore, $x \neq \pm 1$.

5. i **(A)** Since $f(2)$ implies that $x = 2$, $g(f(2)) = 2$. Therefore, $g(f(2)) = 3(f(2)) + 2 = 2$. Therefore, $f(2) = 0$.

6. i **(C)** $p(a) = 0$ implies $4a - 6 = 0$, so $a = \dfrac{3}{2}$.

7. a **(E)** $f \circ g\left(\sqrt{2}\right) = f\left(g\left(\sqrt{2}\right)\right) = f\left(\sin \sqrt{2}\right) = e^{\sin \sqrt{2}} \approx 2.7$.

Inverses

1. i **(E)** If $y = 2x - 3$, the inverse is $x = 2y - 3$, which equals $y = \dfrac{x+3}{2}$.

2. i **(A)** By definition.

3. i **(B)** The inverse is $\{(2,1),(3,2),(4,3),(1,4),(2,5)\}$, which is not a function because of $(2,1)$ and $(2,5)$. Therefore, the domain of the original function must lose either 1 or 5.

4. i **(E)** If this line were reflected about the line $y = x$ to get its inverse, the slope would be less than 1 and the y-intercept would be less than zero. The only possibilities are Choices D and E. Choice D can be excluded because since the x-intercept of $f(x)$ is greater than -1, the y-intercept of its inverse must be greater than -1.

Odd and Even Functions

1. i **(D)** Use the appropriate test for determining whether a relation is even.
   I. The graph of $y = 2$ is a horizontal line, which is symmetric about the y axis, so $y = 2$ is even.
   II. Since $f(-x) = -x \neq x = f(x)$ unless $x = 0$, this function is not even.
   III. Since $(-x)^2 + y^2 = 1$ whenever $x^2 + y^2 = 1$ does, this relation is even.

2. i **(D)** Use the appropriate test for determining whether a relation is odd.
   I. The graph of $y = 2$ is a horizontal line, which is not symmetric about the origin, so $y = 2$ is not odd.
   II. Since $f(-x) = -x = -f(x)$, this function is odd.
   III. Since $(-x)^2 + (-y)^2 = 1$ whenever $x^2 + y^2 = 1$ does, this relation is odd.

3. i **(B)** The analysis of relation III in the above examples indicate that I and II are both even and odd. Since $-x + y \neq 0$ when $x + y = 0$ unless $x = 0$, III is not even, and is therefore not both even and odd.

4. i **(C)** A is even, B is odd, D is even, and E is odd. C is not even because $(-x)^3 - 1 = -x^3 - 1$, which is neither $x^3 - 1$ nor $-x^3 + 1$.

1.2 Polynomial Functions

LINEAR FUNCTIONS

Linear functions are polynomials in which the largest exponent is 1. The graph is always a straight line. Although the general form of the equation is $Ax + By + C = 0$, where A, B, and C are constants, the most useful form occurs when the equation is solved for y. This is known as the *slope-intercept* form and is written $y = mx + b$. The slope of the line is represented by m and is defined to be the ratio of $\dfrac{y_1 - y_2}{x_1 - x_2}$, where (x_1, y_1) and (x_2, y_2) are any two points on the line. The y-intercept is b (the point where the graph crosses the y-axis).

If you solve the general equation of a line, you will find that the slope is $-\dfrac{A}{B}$ and the y-intercept is $-\dfrac{C}{B}$.

You can always quickly write an equation of a line when given its slope and a point on it by using the point-slope form: $y - y_1 = m(x - x_1)$, where m is the slope and (x_1, y_1) is the point. If you are given two points on a line, you must first find the slope using the two points. Then use either point and this slope to write the equation. Once you have the equation in point-slope form, you can always solve for y to get the slope-intercept form if necessary.

Example 1

Write an equation of the line containing (6,–5) and having slope $\dfrac{3}{4}$.

In point-slope form, the equation is $y + 5 = \dfrac{3}{4}(x - 6)$.

Example 2

Write an equation of the line containing (1,–3) and (–4,–2).

First find the slope $m = \dfrac{-3 + 2}{1 + 4} = -\dfrac{1}{5}$. Then use the point (1,–3) and this slope to write the point-slope equation $y + 3 = -\dfrac{1}{5}(x - 1)$.

Parallel lines have the same slope. The slopes of two perpendicular lines are negative reciprocals of one another.

Example 3

The equation of line l_1 is $y = 2x + 3$, and the equation of line l_2 is $y = 2x - 5$.

These lines are parallel because the slope of each line is 2, and the y-intercepts are different.

Example 4

The equation of line l_1 is $y = \dfrac{5}{2}x - 4$, and the equation of line l_2 is $y = -\dfrac{2}{5}x + 9$.

These lines are perpendicular because the slope of l_2, $-\dfrac{2}{5}$, is the negative reciprocal of the slope of l_1, $\dfrac{5}{2}$.

You can use these facts to write an equation of a line that is parallel or perpendicular to a given line and that contains a given point.

Example 5

Write an equation of the line containing (1,7) and parallel to the line $3x + 5y = 8$.

The slope of the given line is $-\dfrac{A}{B} = -\dfrac{3}{5}$. The point-slope equation of the line containing (1,7) is therefore $y - 7 = -\dfrac{3}{5}(x - 1)$.

Example 6

Write an equation of the line containing (–3,2) and perpendicular to $y = 4x - 5$.

The slope of the given line is 4, so the slope of a line perpendicular to it is $-\dfrac{1}{4}$. The desired equation is $y - 2 = -\dfrac{1}{4}(x + 3)$.

The distance between two points P and Q whose coordinates are (x_1, y_1) and (x_2, y_2) is given by the formula

$$\text{Distance} = \sqrt{(x_1 - x_2)^2 + (y_1 - y_2)^2}$$

and the midpoint, M, of the segment $\overline{PQ}$ has coordinates $\left(\dfrac{x_1 + x_2}{2}, \dfrac{y_1 + y_2}{2} \right)$. Graphing calculator programs to find the distances between two points and the midpoint of a segment can be found in Part 3, Chapter 8.

Example 7

Given point (2,–3) and point (–5,4). Find the length of $\overline{PQ}$ and the coordinates of the midpoint, M.

$$PQ = \sqrt{(2 - (-5))^2 + (-3 - 4)^2} \approx 9.9$$

$$M = \left(\frac{2 + (-5)}{2}, \frac{-3 + 4}{2} \right) = \left(\frac{-3}{2}, \frac{1}{2} \right)$$

EXERCISES

1. The slope of the line through points $A(3,-2)$ and $B(-2,-3)$ is

    (A) -5

    (B) $-\dfrac{1}{5}$

    (C) $\dfrac{1}{5}$

    (D) 1

    (E) 5

2. The slope of line $8x + 12y + 5 = 0$ is

    (A) $-\dfrac{3}{2}$

    (B) $-\dfrac{2}{3}$

    (C) $\dfrac{2}{3}$

    (D) 2

    (E) 3

3. The slope of the line perpendicular to line $3x - 5y + 8 = 0$ is

    (A) $-\dfrac{5}{3}$

    (B) $-\dfrac{3}{5}$

    (C) $\dfrac{3}{5}$

    (D) $\dfrac{5}{3}$

    (E) 3

4. The y-intercept of the line through the two points whose coordinates are $(5,-2)$ and $(1,3)$ is

    (A) $-\dfrac{5}{4}$

    (B) $\dfrac{5}{4}$

    (C) $\dfrac{17}{4}$

    (D) 7

    (E) 17

5. The equation of the perpendicular bisector of the segment joining the points whose coordinates are (1,4) and (–2,3) is

   (A) $3x – 2y + 5 = 0$
   (B) $x – 3y + 2 = 0$
   (C) $3x + y – 2 = 0$
   (D) $x – 3y + 11 = 0$
   (E) $x + 3y – 10 = 0$

6. The length of the segment joining the points with coordinates (–2,4) and (3,–5) is

   (A) 2.8
   (B) 3.7
   (C) 10.0
   (D) 10.3
   (E) none of these

7. The slope of the line parallel to the line whose equation is $2x + 3y = 8$ is

   (A) -2
   (B) $-\dfrac{3}{2}$
   (C) $-\dfrac{2}{3}$
   (D) $\dfrac{2}{3}$
   (E) $\dfrac{3}{2}$

8. If the graph of $\pi x + \sqrt{2}\,y + \sqrt{3} = 0$ is perpendicular to the graph of $ax + 3y + 2 = 0$, then $a =$

   (A) -4.5
   (B) -2.22
   (C) -1.35
   (D) 0.45
   (E) 1.35

QUADRATIC FUNCTIONS

Quadratic functions are polynomials in which the largest exponent is 2. The graph is always a parabola. The general form of the equation is $y = ax^2 + bx + c$. If $a > 0$, the parabola opens up and has a minimum value. If $a < 0$, the parabola opens down and has a maximum value. The x-coordinate of the vertex of the parabola is equal to $-\dfrac{b}{2a}$, and the axis of symmetry is the vertical line whose equation is $x = -\dfrac{b}{2a}$.

To find the minimum (or maximum) value of the function, substitute $-\dfrac{b}{2a}$ for x to determine y.

Thus, in the general case the coordinates of the vertex are $\left(-\dfrac{b}{2a}, c - \dfrac{b^2}{4a}\right)$ and the minimum (or maximum) value of the function is $c - \dfrac{b^2}{4a}$.

Unless specifically limited, the domain of a quadratic function is all real numbers, and the range is all values of y greater than or equal to the minimum value (or all values of y less than or equal to the maximum value) of the function.

The examples below provide algebraic underpinnings of how the orientation, vertex, axis of symmetry, and zeros are determined. You can, of course, use a graphing calculator to sketch a parabola and find its vertex and x-intercepts. See Chapter 8.

Example 1

Determine the coordinates of the vertex and the equation of the axis of symmetry of $y = 3x^2 + 2x - 5$. Does the quadratic function have a minimum or maximum value? If so, what is it?

The equation of the axis of symmetry is

$$x = -\frac{b}{2a} = -\frac{2}{2 \cdot 3} = -\frac{1}{3}$$

and the y-coordinate of the vertex is

$$y = 3\left(-\frac{1}{3}\right)^2 + 2\left(-\frac{1}{3}\right) - 5 = -5\frac{1}{3}$$

The vertex is, therefore, at $\left(-\frac{1}{3}, -5\frac{1}{3}\right)$.

The function has a minimum value because $a = 3 > 0$. The minimum value is $-5\frac{1}{3}$. The graph of $y = 3x^2 + 2x - 5$ is shown below.

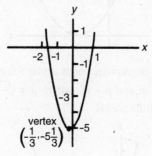

The points where the graph crosses the x-axis are called the *zeros* of the function and occur when $y = 0$. To find the zeros of $y = 3x^2 + 2x - 5$, solve the quadratic equation $3x^2 + 2x - 5 = 0$. The roots of this equation are the zeros of the polynomial.

Every quadratic equation can be changed into the form $ax^2 + bx + c = 0$ (if it is not already in that form), which can be solved by completing the square. The solutions are $x = \dfrac{-b \pm \sqrt{b^2 - 4ac}}{2a}$, the *general quadratic formula*. In the case of $3x^2 + 2x - 5 = 0$, QUADFORM gives solutions 1 and −1.6667. Factoring, $3x^2 + 2x - 5 = (3x + 5)(x - 1) = 0$.

Thus, $3x + 5 = 0$ or $x - 1 = 0$, which leads to $x = -\dfrac{5}{3}$ or 1.

TIP

The graphing calculator program QUADFORM in Chapter 7 provides quick solutions. If the quadratic expression can be factored, this method may be more convenient.

Example 2

Find the zeros of $y = 2x^2 + 3x - 4$.

Solve the equation $2x^2 + 3x - 4 = 0$. This does not factor easily. Using QUADFORM, where $a = 2$, $b = 3$, and $c = -4$, gives $x \approx 0.85078$ and -2.35078.

It is interesting to note that the sum of the two zeros, $\dfrac{-b + \sqrt{b^2 - 4ac}}{2a}$ and $\dfrac{-b - \sqrt{b^2 - 4ac}}{2a}$,

equals $-\dfrac{b}{a}$, and their product equals $\dfrac{c}{a}$. This information can be used to check whether the correct zeros have been found. In Example 2, the sum and product of the zeros can be determined by inspection from the equations. Sum $= -\dfrac{3}{2} = -\dfrac{b}{a}$ and Product $= -1.9999 \approx -2$.

At times it is necessary to determine only the *nature* of the roots of a quadratic equation, not the roots themselves. Because $b^2 - 4ac$ of the general quadratic formula is under the radical, it determines the nature of the roots and is called the *discriminant* of a quadratic equation.

(i) If $b^2 - 4ac = 0$, the two roots are the same $\left(-\dfrac{b}{2a} \right)$, and the graph of the function is tangent to the x-axis.

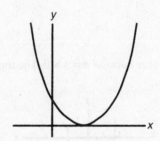

(ii) If $b^2 - 4ac < 0$, there is a negative number under the radical, which gives two complex numbers (of the form $p + qi$ and $p - qi$, where $i = \sqrt{-1}$) as roots, and the graph of the function does not intersect the x-axis.

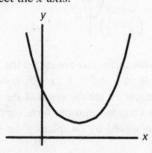

(iii) If $b^2 - 4ac > 0$, there is a positive number under the radical, which gives two different real roots, and the graph of the function intersects the *x*-axis at two points.

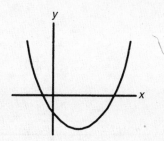

EXERCISES

1. The coordinates of the vertex of the parabola whose equation is $y = 2x^2 + 4x - 5$ are

   (A) $(2, 11)$
   (B) $(-1, -7)$
   (C) $(1, 1)$
   (D) $(-2, -5)$
   (E) $(-4, 11)$

2. The range of the function
   $f = \{(x,y):y = 5 - 4x - x^2\}$ is

   (A) $\{y:y \le 0\}$
   (B) $\{y:y \ge -9\}$
   (C) $\{y:y \le 9\}$
   (D) $\{y:y \ge 0\}$
   (E) $\{y:y \le 1\}$

3. The equation of the axis of symmetry of the function $y = 2x^2 + 3x - 6$ is

   (A) $x = -\dfrac{3}{2}$

   (B) $x = -\dfrac{3}{4}$

   (C) $x = -\dfrac{1}{3}$

   (D) $x = \dfrac{1}{3}$

   (E) $x = \dfrac{3}{4}$

4. Find the zeros of $y = 2x^2 + x - 6$.

   (A) 3 and 2

   (B) -3 and 2

   (C) $\dfrac{1}{2}$ and $\dfrac{3}{2}$

   (D) $-\dfrac{3}{2}$ and 1

   (E) $\dfrac{3}{2}$ and -2

5. The sum of the zeros of $y = 3x^2 - 6x - 4$ is

   (A) -2

   (B) $-\dfrac{4}{3}$

   (C) $\dfrac{4}{3}$

   (D) 2

   (E) 6

6. $x^2 + 2x + 3 = 0$ has

   (A) two real rational roots
   (B) two real irrational roots
   (C) two equal real roots
   (D) two equal rational roots
   (E) two complex conjugate roots

7. A parabola with a vertical axis has its vertex at the origin and passes through point (7,7). The parabola intersects line $y = 6$ at two points. The length of the segment joining these points is

   (A) 14
   (B) 13
   (C) 12
   (D) 8.6
   (E) 6.5

HIGHER-DEGREE POLYNOMIAL FUNCTIONS

Polynomial functions of degree greater than two (largest exponent greater than 2) are usually treated together since there are no simple formulas, such as the general quadratic formula, that aid in finding zeros.

Five facts about the graphs of polynomial functions:

1. They are always continuous curves. (The graph can be drawn without removing the pencil from the paper.)
2. If the largest exponent is an even number, both ends of the graph leave the coordinate system either at the top or at the bottom:

Example 1

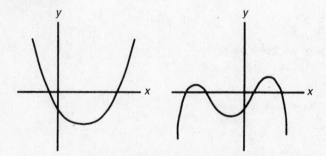

3. If the largest exponent is an odd number, the ends of the graph leave the coordinate system at opposite ends.

Example 2

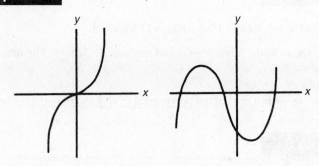

Facts (2) and (3) describe the *end behavior* of a polynomial.

4. If all the exponents are even numbers, the polynomial is an *even function* and therefore symmetric about the *y*-axis.

Example 3

$$y = 3x^4 + 2x^2 - 8$$

5. If all the exponents are odd numbers and there is no constant term, the polynomial is an *odd function* and therefore symmetric about the origin of the coordinate system.

Example 4

$$y = 4x^5 + 2x^3 - 3x$$

A polynomial of degree n has n zeros. The zeros of polynomials with real coefficients can be real or imaginary numbers, but imaginary zeros must occur in pairs. For example, if the degree of a polynomial is 6, there are 6 real zeros and no imaginary zeros; 4 real zeros and 2 imaginary ones; 2 real zeros and 4 imaginary ones, or 6 imaginary zeros. Moreover, real zeros can occur more than once.

If a real zero occurs n times, it is said to have multiplicity n. If a zero of a polynomial has even multiplicity, its graph touches, but does not cross the *x*-axis. For example, in the polynomial $f(x) = (x - 3)^4$, 3 is a zero of multiplicity 4, and the graph of $f(x)$ is tangent to the *x*-axis at $x = 3$. If a zero of a polynomial has odd multiplicity, its graph crosses the *x*-axis. For example, in the polynomial $f(x) = (x - 1)^5$, 1 is a zero of multiplicity 5, and the graph of $f(x)$ crosses the *x*-axis at $x = 1$.

The multiplicity of a real zero is counted toward the total number of zeros. For example, the polynomial $f(x) = (x - 5)(x + 3)^4(x^2 + 6)$ has degree 7: one zero (5) of multiplicity 1; one zero (−3) of multiplicity 4; and two imaginary zeros ($i\sqrt{6}$ and $-i\sqrt{6}$).

There are 5 facts that are useful when analyzing polynomial functions.

1. **Remainder theorem**—If a polynomial $P(x)$ is divided by $x - r$ (where r is any constant), then the remainder is $P(r)$.

Example 5

Divide $P(x) = 3x^5 - 4x^4 - 15x^2 - 88x - 12$ by $x - 3$.

Enter $P(x)$ into Y_1, return to the home screen, and evaluate $Y_1(3) = -6$. The remainder is -6 when you divide $P(x)$ by $x - 3$.

2. **Factor theorem**—r is a zero of the polynomial $P(x)$ if and only if $x - r$ is a divisor of $P(x)$.

Example 6

Is $x - 99$ a factor of $x^4 - 100x^3 + 97x^2 + 200x - 198$?

Call this polynomial $P(x)$ and evaluate $P(99)$ using your graphing calculator. Since $P(99) = 0$, the answer to the question is "Yes."

3. **Rational zero (root) theorem**—If $\dfrac{p}{q}$ is a rational zero (reduced to lowest terms) of a polynomial $P(x)$ with integral coefficients, then p is a factor of a_0 (the constant term) and q is a factor of a_n (the leading coefficient).

Example 7

What are the possible rational zeros of $P(x) = 3x^3 + 2x^2 + 4x - 6$?

The divisors of the constant term -6 are ±1, ±2, ±3, ±6, and the divisors of the leading coefficient 3 are ±1, ±3. The 12 rational numbers that could possibly be zeros of $P(x)$ are ±1, ±2, ±3, ±6, $\pm\dfrac{1}{3}$, $\pm\dfrac{2}{3}$, the ratios of all numbers in the first group to all numbers in the second.

4. If $P(x)$ is a polynomial with real coefficients, then complex zeros occur as conjugate pairs. (For example, if $p + qi$ is a zero, then $p - qi$ is also a zero.)

Example 8

If $3 + 2i$, 2, and $2 - 3i$ are all zeros of $P(x) = 3x^5 - 36x^4 + 2x^3 - 8x^2 + 9x - 338$, what are the other zeros?

Since $P(x)$ must have five zeros because it is a fifth-degree polynomial, and since the coefficients are real and the complex zeros come in conjugate pairs, the two remaining zeros must be $3 - 2i$ and $2 + 3i$.

5. **Descartes' rule of signs**—The number of positive real zeros of a polynomial $P(x)$ either is equal to the number of variations in the sign between terms or is less than that number by an even integer. The number of negative real zeros of $P(x)$ either is equal to the number of variations of the sign between the terms of $P(-x)$ or is less than that number by an even integer.

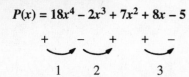

$P(x) = 18x^4 - 2x^3 + 7x^2 + 8x - 5$

$$\underbrace{+ \quad -}_{1} \quad \underbrace{- \quad +}_{2} \quad \underbrace{+ \quad + \quad -}_{3}$$

Three sign changes indicate there will be either one or three positive zeros of $P(x)$.

$$P(-x) = 18x^4 + 2x^3 + 7x^2 - 8x - 5$$

$$+ \quad + \quad \underbrace{+ \quad -}_{1} \quad -$$

One sign change indicates there will be exactly one negative zero of $P(x)$.

EXERCISES

1. $P(x) = ax^4 + x^3 - bx^2 - 4x + c$. If $P(x)$ increases without bound as x increases without bound, then, as x decreases without bound, $P(x)$

   (A) increases without bound
   (B) decreases without bound
   (C) approaches zero from above the x-axis
   (D) approaches zero from below the x-axis
   (E) cannot be determined

2. Which of the following is an odd function?

   I. $f(x) = 3x^3 + 5$
   II. $g(x) = 4x^6 + 2x^4 - 3x^2$
   III. $h(x) = 7x^5 - 8x^3 + 12x$

   (A) only I
   (B) only II
   (C) only III
   (D) only I and II
   (E) only I and III

3. How many possible rational roots are there for $2x^4 + 4x^3 - 6x^2 + 15x - 12 = 0$?

   (A) 4
   (B) 6
   (C) 8
   (D) 12
   (E) 16

4. If both $x - 1$ and $x - 2$ are factors of $x^3 - 3x^2 + 2x - 4b$, then b must be

   (A) 0
   (B) 1
   (C) 2
   (D) 3
   (E) 4

5. If $3x^3 - 9x^2 + Kx - 12$ is divisible by $x - 3$, then $K =$

(A) -40
(B) -3
(C) 3
(D) 4
(E) 22

6. Write the equation of lowest degree with real coefficients if two of its roots are -1 and $1 + i$.

(A) $x^3 + x^2 + 2 = 0$
(B) $x^3 - x^2 - 2 = 0$
(C) $x^3 - x + 2 = 0$
(D) $x^3 - x^2 + 2 = 0$
(E) none of the above

INEQUALITIES

Given any algebraic expression $f(x)$, there are exactly three situations that can exist:

1. for some values of x, $f(x) < 0$;
2. for some values of x, $f(x) = 0$;
3. for some values of x, $f(x) > 0$.

If all three of these sets of numbers are indicated on a number line, the set of values that satisfy $f(x) < 0$ is always separated from the set of values that satisfy $f(x) > 0$ by the values of x that satisfy $f(x) = 0$.

Example

Find the set of values for x that satisfies $x^2 - 3x - 4 < 0$.

Graph $y = x^2 - 3x - 4$. You need to find the x values of points on the graph that lie below the x-axis. First find the zeros: $x = 4$, $x = -1$. The points that lie below the x-axis are (strictly) between -1 and 4, or $-1 < x < 4$.

EXERCISES

1. Which of the following is equivalent to $3x^2 - x < 2$?

(A) $-\dfrac{3}{2} < x < 1$

(B) $-1 < x < \dfrac{2}{3}$

(C) $-\dfrac{2}{3} < x < 1$

(D) $-1 < x < \dfrac{3}{2}$

(E) $x < -\dfrac{2}{3}$ or $x > 1$

2. Solve $x^5 - 3x^3 + 2x^2 - 3 > 0$.

    (A) $(-\infty, -0.87)$
    (B) $(-1.90, -0.87)$
    (C) $(-1.90, -0.87) \cup (1.58, \infty)$
    (D) $(-0.87, 1.58)$
    (E) $(1.58, \infty)$

3. The number of integers that satisfy the inequality $x^2 + 48 < 16x$ is

    (A) 0
    (B) 4
    (C) 7
    (D) an infinite number
    (E) none of the above

Answers and Explanations

In these solutions the following notation is used:

    i: calculator unnecessary
    a: calculator helpful or necessary
    g: graphing calculator helpful or necessary

Linear Functions

1. i   **(C)** Slope $= \dfrac{-3-(-2)}{-2-3} = \dfrac{1}{5}$.

2. i   **(B)** $y = -\dfrac{2}{3}x - \dfrac{5}{12}$. The slope is $-\dfrac{2}{3}$.

3. i   **(A)** $y = \dfrac{3}{5}x + \dfrac{8}{5}$. The slope of the given line is $\dfrac{3}{5}$. The slope of a perpendicular line

    is $-\dfrac{5}{3}$.

4. i   **(C)** The slope of the line is $\dfrac{-2-3}{5-1} = -\dfrac{5}{4}$, so the point-slope equation is

    $y - 3 = -\dfrac{5}{4}(x - 1)$. Solve for y to get $y = -\dfrac{5}{4}x + \dfrac{17}{4}$. The y-intercept of the line is $\dfrac{17}{4}$.

5. i   **(C)** The slope of the segment is $\dfrac{3-4}{-2-1} = \dfrac{1}{3}$. Therefore, the slope of a perpendicular

    line is -3. The midpoint of the segment is $\left(\dfrac{1-2}{2}, \dfrac{4+3}{2}\right) = \left(-\dfrac{1}{2}, \dfrac{7}{2}\right)$. Therefore, the

    point-slope equation is $y - \dfrac{7}{2} = -3\left(x + \dfrac{1}{2}\right)$. In general form, this equation is

    $3x + y - 2 = 0$.

6. a **(D)** Length $= \sqrt{(3+2)^2 + (-5-4)^2} \approx 10.3$

7. i **(C)** $y = -\dfrac{2}{3}x + \dfrac{8}{3}$. Therefore, the slope of a parallel line $= -\dfrac{2}{3}$.

8. a **(C)** The slope of the first line is $-\dfrac{\pi}{\sqrt{2}}$, and the slope of the second line is $-\dfrac{a}{3}$. To be

   perpendicular, $-\dfrac{\pi}{\sqrt{2}} = \dfrac{3}{a}$. $a = \dfrac{-3\sqrt{2}}{\pi} \approx -1.35$.

Quadratic Functions

1. i **(B)** The x coordinate of the vertex is $x = -\dfrac{b}{2a} = -\dfrac{4}{4} = -1$ and the y coordinate is

   $y = 2(-1)^2 + 4(-1) - 5 = -7$. Hence the vertex is the point $(-1,-7)$.

2. i **(C)** Find the vertex: $x = -\dfrac{b}{2a} = \dfrac{4}{-2} = -2$ and $y = 5 - 4(-2) - (-2)^2 = 9$. Since

   $a = -1 < 0$ the parabola opens down, so the range is $\{y : y \leq 9\}$.

3. i **(B)** The x coordinate of the vertex is $x = -\dfrac{b}{2a} = -\dfrac{3}{4}$. Thus, the equation of the axis

   of symmetry is $x = -\dfrac{3}{4}$.

4. i **(E)** $2x^2 + x - 6 = (2x - 3)(x + 2) = 0$. The zeros are $\dfrac{3}{2}$ and -2.

5. i **(D)** Sum of zeros $= -\dfrac{b}{a} = -\dfrac{-6}{3} = 2$.

6. i **(E)** From the discriminant $b^2 - 4ac = 4 - 4 \cdot 1 \cdot 3 = -8 < 0$.

7. a **(B)** The equation of a vertical parabola with its vertex at the origin has the form $y = ax^2$.

   Substitute $(7,7)$ for x and y to find $a = \dfrac{1}{7}$. When $y = 6$, $x^2 = 42$. Therefore, $x = \pm\sqrt{42}$,

   and the segment $= 2\sqrt{42} \approx 13$.

Higher-Degree Polynomial Functions

1. i **(A)** Since the degree of the polynomial is an even number, both ends of the graph go off in the same direction. Since $P(x)$ increases without bound as x increases, $P(x)$ also increases without bound as x decreases.

2. i **(C)** Since the exponents are all odd, and there is no constant term, III is the only odd function.

3. i **(E)** Rational roots have the form $\dfrac{p}{q}$, where p is a factor of 12 and q is a factor of 2.

$$\frac{p}{q} \in \left\{\pm 12, \pm 6, \pm 4, \pm 3, \pm 2, \pm 1, \pm\frac{3}{2}, \pm\frac{1}{2}\right\}. \text{ The total is } 16.$$

4. a **(A)** Since $x - 1$ is a factor, $P(1) = 1^3 - 3 \cdot 1^2 + 2 \cdot 1 - 4b = 0$. Therefore, $b = 0$.

5. a **(D)** Substitute 3 for x set equal to zero and solve for K.

6. i **(D)** $1 - i$ is also a root. To find the equation, multiply $(x + 1)[x - (1 + i)][x - (1 - i)]$, which are the factors that produced the three roots.

Inequalities

1. i **(C)** $3x^2 - x - 2 = (3x + 2)(x - 1) = 0$ when $x = -\dfrac{2}{3}$ or 1. Numbers between these satisfy the original inequality.

2. g **(D)** Graph the function, and determine that the three zeros are -1.90, -0.87, and 1.58. The parts of the graph that are above the x-axis have x-coordinates between -1.90 and -0.87 and are larger than 1.58.

3. i **(C)** $x^2 - 16x + 48 = (x - 4)(x - 12) = 0$, when $x = 4$ or 12. Numbers between these satisfy the original inequality.

1.3 Trigonometric Functions and Their Inverses

DEFINITIONS

The general definitions of the six trigonometric functions are obtained from an angle placed in standard position on a rectangular coordinate system. When an angle θ is placed so that its vertex is at the origin, its initial side is along the positive x-axis, and its terminal side is anywhere on the coordinate system, it is said to be in *standard position*. The angle is given a positive value if it is measured in a counterclockwise direction from the initial side to the terminal side, and a negative value if it is measured in a clockwise direction.

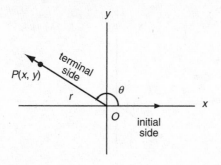

Let $P(x,y)$ be any point on the terminal side of the angle, and let r represent the distance between O and P. The six trigonometric functions are defined to be:

$$\sin \theta = \frac{y}{r}$$

$$\cos \theta = \frac{x}{r}$$

$$\tan \theta = \frac{y}{x}$$

$$\cot \theta = \frac{x}{y}$$

$$\sec \theta = \frac{r}{x}$$

$$\csc \theta = \frac{r}{y}$$

> **ALERT!**
>
> $\sin \theta$ and $\cos \theta$ are always between -1 and 1.

From these definitions it follows that:

$$\sin \theta \cdot \csc \theta = 1 \qquad \tan \theta = \frac{\sin \theta}{\cos \theta}$$

$$\cos \theta \cdot \sec \theta = 1 \qquad \cot \theta = \frac{\cos \theta}{\sin \theta}$$

$$\tan \theta \cdot \cot \theta = 1$$

The distance OP is always positive, and the x and y coordinates of P are positive or negative depending on which quadrant the terminal side of $\angle \theta$ lies in. The signs of the trigonometric functions are indicated in the following table.

> **TIP**
>
> **A**ll trig functions are positive in quadrant I.
>
> **S**ine and only sine is positive in quadrant II.
>
> **T**angent and only tangent is positive in quadrant III.
>
> **C**osine and only cosine is positive in quadrant IV.
>
> Just remember: **A**ll **S**tudents **T**ake **C**alculus.

Quadrant	I	II	III	IV
Function: $\sin \theta$, $\csc \theta$	+	+	−	−
$\cos \theta$, $\sec \theta$	+	−	−	+
$\tan \theta$, $\cot \theta$	+	−	+	−

Each angle θ whose terminal side lies in quadrant II, III, or IV has associated with it an angle called the *reference angle;* $\angle \alpha$ is formed by the x-axis and the terminal side.

Any trig function of $\angle \theta = \pm$ the same function of $\angle \alpha$. The sign is determined by the quadrant in which the terminal side lies.

Example 1

Express sin 320° in terms of ∠α.

$$\alpha = 360° - 320° = 40°$$

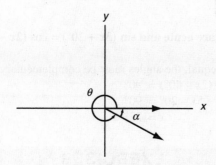

Since the sine is negative in quadrant IV, sin 320° = –sin 40°.

Example 2

Express cot 200° in terms of ∠α.

$$\alpha = 200° - 180° = 20°$$

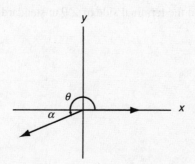

Since the cotangent is positive in quadrant III, cot 200° = cot 20°.

Example 3

Express cos 130° in terms of ∠α.

$$\alpha = 180° - 130° = 50°$$

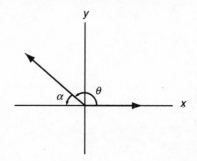

Since the cosine is negative in quadrant II, cos 130° = –cos 50°.

Sine and cosine, tangent and cotangent, and secant and cosecant are *cofunction pairs*. These examples indicate that any function of α = the cofunction of β and that α and β are *complementary* angles because their sum is 90°. A very useful property is obtained from this observation: *Cofunctions of complementary angles are equal.*

Example 4

If both the angles are acute and sin $(3x + 20°)$ = cos $(2x - 40°)$, find x.

Since these cofunctions are equal, the angles must be complementary.

Therefore, $(3x + 20°) + (2x - 40°) = 90°$

$$5x - 20° = 90°$$
$$x = 22°$$

EXERCISES

1. Express cos 320° as a function of an angle between 0° and 90°.

    (A) cos 40°
    (B) sin 40°
    (C) cos 50°
    (D) sin 50°
    (E) none of the above

2. If point $P(-5,12)$ lies on the terminal side of $\angle\theta$ in standard position, sin $\theta =$

    (A) $-\dfrac{12}{13}$

    (B) $\dfrac{-5}{12}$

    (C) $\dfrac{-5}{13}$

    (D) $\dfrac{12}{13}$

    (E) $\dfrac{12}{5}$

3. If sec $\theta = -\dfrac{5}{4}$ and sin $\theta > 0$, then tan $\theta =$

    (A) $\dfrac{4}{3}$

    (B) $\dfrac{3}{4}$

    (C) $-\dfrac{3}{4}$

    (D) $-\dfrac{4}{3}$

    (E) none of the above

4. If x is an angle in quadrant III and $\tan (x - 30°) = \cot x$, find x.

(A) 240°
(B) 225°
(C) 210°
(D) 60°
(E) none of the above

5. If $90° < \alpha < 180°$ and $270° < \beta < 360°$, then which of the following *cannot* be true?

(A) $\sin \alpha = \sin \beta$
(B) $\tan \alpha = \sin \beta$
(C) $\tan \alpha = \tan \beta$
(D) $\sin \alpha = \cos \beta$
(E) $\sec \alpha = \csc \beta$

6. Expressed as a function of an acute angle, $\cos 310° + \cos 190° =$

(A) $-\cos 40°$
(B) $\cos 70°$
(C) $-\cos 50°$
(D) $\sin 20°$
(E) $-\cos 70°$

ARCS AND ANGLES

Although the degree is the chief unit used to measure an angle in elementary mathematics courses, the radian has several advantages in more advanced mathematics. Degrees and radians are related by this equation: $\pi^R = 180°$.

> **TIP**
>
> Although ^{R} is used to indicate radians, a radian actually has no units, so the use of ^{R} is optional.

Example 1

In each of the following, convert the degrees to radians or the radians to degrees. (If no unit of measurement is indicated, radians are assumed.)

(A) 30°

(B) 270°

(C) $\dfrac{\pi}{4}$

(D) $\dfrac{17\pi}{3}$

(E) 24

SOLUTIONS

(A) To change degrees to radians multiply by $\dfrac{\pi}{180°}$, so $30° = 30°\left(\dfrac{\pi}{180°}\right) = \dfrac{\pi^R}{6}$.

(B) $270°\left(\dfrac{\pi}{180°}\right) = \dfrac{3\pi^R}{2}$

(C) To change radians to degrees, multiply by $\dfrac{180°}{\pi^R}$, so $\dfrac{\pi}{4} \cdot \dfrac{180°}{\pi^R} = 45°$

(D) $\dfrac{17\pi^{R}}{3} \cdot \dfrac{180°}{\pi^{R}} = 1020°$

(E) $24^{R}\left(\dfrac{180°}{\pi^{R}}\right) = \left(\dfrac{4320}{\pi}\right)° \approx 1375°$

In a circle of radius r inches with an arc subtended by a central angle of θ measured in radians, two important formulas can be derived. The length of the arc, s, is equal to $r\theta$, and the area of the sector, AOB, is equal to $\dfrac{1}{2}r^{2}\theta$.

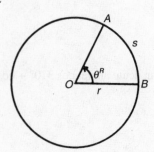

Example 2

Find the area of the sector and the length of the arc subtended by a central angle of $\dfrac{2\pi}{3}$ radians in a circle whose radius is 6 inches.

$s = r\theta$ $\qquad\qquad$ $A = \dfrac{1}{2}r^{2}\theta$

$s = 6 \cdot \dfrac{2\pi}{3} = 4\pi$ inches $\qquad$ $A = \dfrac{1}{2} \cdot 36 \cdot \dfrac{2\pi}{3} = 12\pi$ square inches

Example 3

In a circle of radius 8 inches, find the area of the sector whose arc length is 6π inches.

$s = r\theta$ $\qquad\qquad$ $A = \dfrac{1}{2}r^{2}\theta$

$6\pi = 8\theta$ $\qquad\qquad$ $A = \dfrac{1}{2} \cdot 64 \cdot \dfrac{3\pi}{4} = 24\pi$ square inches

$\theta = \dfrac{3\pi^{R}}{4}$

Example 4

Find the length of the radius of a circle in which a central angle of 60° subtends an arc of length 8π inches.

The 60° angle must be converted to radians:

$$60° = 60°\left(\frac{\pi}{180°}\right) \text{ radians} = \frac{\pi}{3} \text{ radians}$$

Therefore,

$$s = r\theta$$

$$8\pi = r \cdot \frac{\pi}{3}$$

$$r = 24 \text{ inches}$$

EXERCISES

1. An angle of 30 radians is equal to how many degrees?

   (A) $\dfrac{\pi}{30}$

   (B) $\dfrac{\pi}{6}$

   (C) $\dfrac{30}{\pi}$

   (D) $\dfrac{540}{\pi}$

   (E) $\dfrac{5400}{\pi}$

2. If a sector of a circle has an arc length of 2π inches and an area of 6π square inches, what is the length of the radius of the circle?

   (A) 1
   (B) 2
   (C) 3
   (D) 6
   (E) 12

3. If a circle has a circumference of 16 inches, the area of a sector with a central angle of 4.7 radians is

   (A) 10
   (B) 12
   (C) 15
   (D) 25
   (E) 48

4. A central angle of 40° in a circle of radius 1 inch intercepts an arc whose length is *s*. Find *s*.

    (A) 0.7
    (B) 1.4
    (C) 2.0
    (D) 3.0
    (E) 40

5. The pendulum on a clock swings through an angle of 25°, and the tip sweeps out an arc of 12 inches. How long is the pendulum?

    (A) 1.67 inches
    (B) 13.8 inches
    (C) 27.5 inches
    (D) 43.2 inches
    (E) 86.4 inches

SPECIAL ANGLES

When you use a calculator to evaluate most trig values, you will get a decimal approximation. You can use your knowledge of the definitions of the trigonometric functions, reference angles, and the ratios of the sides of the 45°-45°-90° triangle and the 30°-60°-90° triangle ("special" triangles) to get exact trig values for "special" angles: multiples of $30° \left(\dfrac{\pi}{6} \right)$, $45° \left(\dfrac{\pi}{4} \right)$, $60° \left(\dfrac{\pi}{3} \right)$.

The ratios of the sides of the two special triangles are shown in the figure below.

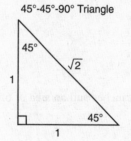

To illustrate how this can be done, suppose you want to find the trig values of $120° \left(\dfrac{2\pi}{3} \right)$. First sketch the following graph.

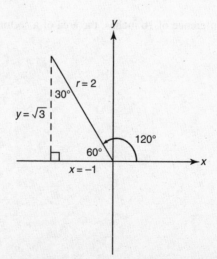

The graph shows the angle in standard position, the reference angle 60°, and the (signed) side length ratios for the 30°-60°-90° triangle. You can now use the definitions of the trig functions to find the trig values:

$$\sin 120° = \frac{y}{r} = \frac{\sqrt{3}}{2} \qquad \cos 120° = \frac{x}{r} = \frac{-1}{2} \qquad \tan 120° = \frac{y}{x} = \frac{\sqrt{3}}{-1} = -\sqrt{3}$$

$$\csc 120° = \frac{r}{y} = \frac{2}{\sqrt{3}} = \frac{2\sqrt{3}}{3} \qquad \sec 120° = \frac{r}{x} = \frac{2}{-1} = -2 \qquad \cot 120° = \frac{x}{y} = \frac{-1}{\sqrt{3}} = -\frac{\sqrt{3}}{3}$$

Values can be checked by comparing the decimal approximation the calculator provides for the trig function with the decimal approximation obtained by entering the exact value in a calculator. In this exmaple, $\sin 120° \approx 0.866$ and $\frac{\sqrt{3}}{2} \approx 0.866$.

You can also readily obtain trig values of the quadrantal angles—multiples of $90° \left(\frac{\pi}{2}\right)$. The terminal sides of these angles are the *x*- and *y*-axes. In these cases, you don't have a triangle at all; instead, either *x* or *y* equals 1 or –1, the other coordinate equals zero, and *r* equals 1. To illustrate how to use this method to evaluate the trig values of 270°, first draw the figure below.

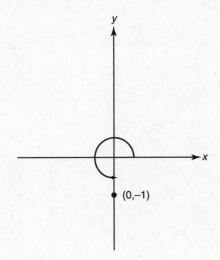

The figure indicates $x = 0$ and $y = -1$ ($r = 1$). Therefore,

$$\sin 270° = \frac{-1}{1} = -1 \qquad \cos 270° = \frac{0}{1} = 0 \qquad \tan 270° = \frac{-1}{0}, \text{ which is undefined}$$

$$\csc 270° = \frac{1}{-1} = -1 \qquad \sec 270° = \frac{1}{0}, \text{ which is undefined} \qquad \cot 270° = \frac{0}{-1} = 0$$

1. The exact value of tan (–60°) is

   (A) $-\sqrt{3}$

   (B) -1

   (C) $-\dfrac{2}{\sqrt{3}}$

   (D) $-\dfrac{\sqrt{3}}{2}$

   (E) $-\dfrac{1}{\sqrt{3}}$

2. The exact value of $\cos \dfrac{3\pi}{4}$

   (A) -1

   (B) $-\dfrac{\sqrt{3}}{2}$

   (C) $-\dfrac{\sqrt{2}}{2}$

   (D) $-\dfrac{1}{2}$

   (E) 0

3. Csc 540° is

   (A) 0

   (B) $-\sqrt{3}$

   (C) $-\sqrt{2}$

   (D) -1

   (E) undefined

GRAPHS

Analyzing the graph of a trigonometric function can be readily accomplished with the aid of a graphing calculator. Such an analysis can determine the amplitude, maximum, minimum, period, or phase shift of a trig function, or solve a trig equation or inequality. Graphing calculator methods for performing these analyses can be found in Chapter 7.

The examples and exercises in this and the next two sections show how a variety of trig problems can be solved without using a graphing calculator. They also explain how to solve trig equations and inequalities and how to analyze inverse trig functions.

Since the values of all the trigonometric functions repeat themselves at regular intervals, and, for some number p, $f(x) = f(x + p)$ for all numbers x, these functions are called *periodic functions*. The smallest positive value of p for which this property holds is called the *period* of the function.

The sine, cosine, secant, and cosecant have periods of 2π, and the tangent and cotangent have periods of π. The graphs of the six trigonometric functions, shown below, demonstrate that the tangent and cotangent repeat on intervals of length π and that the others repeat on intervals of length 2π.

The domain and range of each of the six trigonometric functions are summarized in the table.

	DOMAIN	RANGE
sine	all real numbers	$-1 \le \sin x \le 1$
cosine	all real numbers	$-1 \le \cos x \le 1$
tangent	all real numbers except odd multiples of $\dfrac{\pi}{2}$	all real numbers
cotangent	all real numbers except all multiples of π	all real numbers
secant	all real numbers except odd multiples of $\dfrac{\pi}{2}$	$\sec x \le -1$ or $\sec x \ge 1$
cosecant	all real numbers except all multiples of π	$\csc x \le -1$ or $\csc x \ge 1$

The general form of a trigonometric function, f, is given to be $y = A \cdot f(Bx + C)$, where $|A|$ is the amplitude, $\dfrac{\text{normal period of } f}{B}$ is the period of the graph, and $-\dfrac{C}{B}$ is the phase shift (horizontal translations). These transformations are normally applied to the sine and cosine functions. Occasionally, period and phase shift transformations are applied to the tangent. The *amplitude* is one-half the vertical distance between the lowest and highest point of the graph. The *phase shift* is the distance to the right or left that the graph is changed from its normal position.

The *frequency* is equal to $\dfrac{1}{\text{period}}$.

> **NOTE**
>
> $\sin x$ is an odd function.

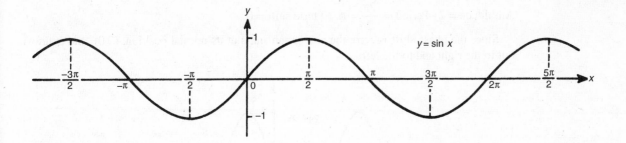

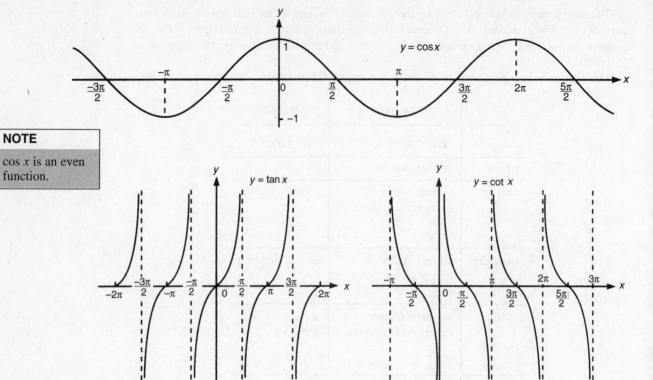

Example 1

Determine the amplitude, period, and phase shift of $y = 2\sin 2x$ and sketch at least one period of the graph.

Amplitude = 2   Period = $\dfrac{2\pi}{2} = \pi$ Phase shift = 0

Since the phase shift is zero, the sine graph starts at its normal position, (0,0), and is drawn out to the right and to the left.

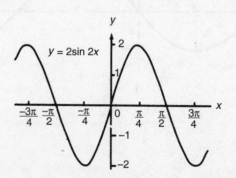

Example 2

Determine the amplitude, period, and phase shift of $y = \frac{1}{2}\cos\left(\frac{1}{2}x - \frac{\pi}{3}\right)$ and sketch at least one period of the graph.

Although a graphing calculator can be used to determine the amplitude, period, and phase shift of a periodic function, it may be more efficient to derive them directly from the equation.

Amplitude $= \dfrac{1}{2}$ Period $= \dfrac{2\pi}{\frac{1}{2}} = 4\pi$

Phase shift $= \dfrac{\frac{\pi}{3}}{\frac{1}{2}} = \dfrac{2\pi}{3}$

Since the phase shift is $\dfrac{2\pi}{3}$, the cosine graph starts at $x = \dfrac{2\pi}{3}$ instead of $x = 0$ and one

period ends at $x = \dfrac{2\pi}{3} + 4\pi$ or $\dfrac{14\pi}{3}$.

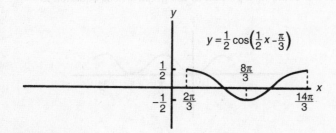

Example 3

Determine the amplitude, period, and phase shift of $y = -2\sin(\pi x + 3\pi)$ and sketch at least one period of the graph.

Amplitude $= 2$ Period $= \dfrac{2\pi}{\pi} = 2$

Phase shift $= -\dfrac{3\pi}{\pi} = -3$

Since the phase shift is -3, the sine graph starts at $x = -3$ instead of $x = 0$, and one period ends at $-3 + 2$ or $x = -1$. The graph can continue to the right and to the left for as many periods as desired. Since the coefficient of the sine is negative, the graph starts down as x increases from -3, instead of up as a normal sine graph does.

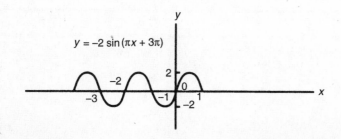

EXERCISES

1. In the figure, part of the graph of $y = \sin 2x$ is shown. What are the coordinates of point P?

    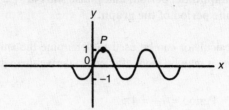

    (A) $\left(\dfrac{\pi}{2}, 1\right)$

    (B) $(\pi, 1)$

    (C) $\left(\dfrac{\pi}{4}, 1\right)$

    (D) $\left(\dfrac{\pi}{2}, 2\right)$

    (E) $(\pi, 2)$

2. The figure below could be a portion of the graph whose equation is

    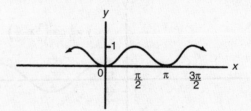

    (A) $y - 1 = \sin x \cdot \cos x$
    (B) $y \sec x = 1$
    (C) $2y + 1 = \sin 2x$
    (D) $2y + 1 = \cos 2x$
    (E) $1 - 2y = \cos 2x$

3. As θ increases from $\dfrac{\pi}{4}$ to $\dfrac{5\pi}{4}$, the value of $4 \cos \dfrac{1}{2}\theta$

    (A) increases, and then decreases
    (B) decreases, and then increases
    (C) decreases throughout
    (D) increases throughout
    (E) decreases, increases, and then decreases again

4. The function $f(x) = \sqrt{3} \cos x + \sin x$ has an amplitude of

    (A) 1.37
    (B) 1.73
    (C) 2
    (D) 2.73
    (E) 3.46

5. For what value of P is the period of the function $y = \frac{1}{3}\cos Px$ equal to $\frac{2\pi}{3}$?

    (A) $\frac{1}{3}$

    (B) $\frac{2}{3}$

    (C) 2

    (D) 3

    (E) 6

6. If $0 \le x \le \frac{\pi}{2}$, what is the maximum value of the function $f(x) = \sin\frac{1}{3}x$?

    (A) 0

    (B) $\frac{1}{3}$

    (C) $\frac{1}{2}$

    (D) $\frac{\sqrt{3}}{2}$

    (E) 1

7. If the graph in the figure below has an equation of the form $y = \sin(Mx + N)$, what is the value of N?

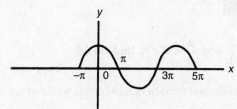

    (A) $-\pi$

    (B) -1

    (C) $-\frac{1}{2}$

    (D) $\frac{\pi}{2}$

    (E) π

IDENTITIES, EQUATIONS, AND INEQUALITIES

There are a few trigonometric identities you must know for the Mathematics Level 2 Subject Test.

- **Reciprocal Identities** recognize the definitional relationships:

$$\csc x = \frac{1}{\sin x} \qquad \sec x = \frac{1}{\cos x} \qquad \cot x = \frac{1}{\tan x}$$

- **Cofunction Identities** were discussed earlier. Using radian measure:

$$\sin x = \cos\left(\frac{\pi}{2} - x\right) \text{ and } \cos x = \sin\left(\frac{\pi}{2} - x\right)$$

$$\sec x = \csc\left(\frac{\pi}{2} - x\right) \text{ and } \csc x = \sec\left(\frac{\pi}{2} - x\right)$$

$$\tan x = \cot\left(\frac{\pi}{2} - x\right) \text{ and } \cot x = \tan\left(\frac{\pi}{2} - x\right)$$

- **Pythagorean Identities**

$$\sin^2 x + \cos^2 x = 1 \qquad \sec^2 x = 1 + \tan^2 x \qquad \csc^2 x = 1 + \cot^2 x$$

- **Double Angle Formulas**

$$\sin 2x = 2(\sin x)(\cos x)$$
$$\cos 2x = \cos^2 x - \sin^2 x$$
$$= 2\cos^2 x - 1$$
$$= 1 - 2\sin^2 x$$

Example 1

Given $\cos \theta = -\dfrac{2}{3}$ and $\dfrac{\pi}{2} < \theta < \pi$, find $\sin 2\theta$.

Since $\sin 2\theta = 2(\sin \theta)(\cos \theta)$, you need to determine the value of $\sin \theta$. From the figure below, you can see that $\sin \theta = \dfrac{\sqrt{5}}{2}$. Therefore, $\sin 2\theta = 2\,\dfrac{\sqrt{5}}{3}\left(-\dfrac{2}{3}\right) = -\dfrac{4}{9}\sqrt{5}$.

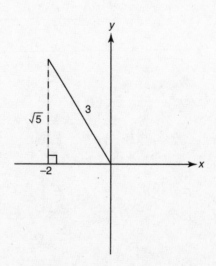

Example 2

If cos 23° = z, find the value of cos 46° in terms of z.

Since $46 = 2(23)$, Formula 12 can be used: $\cos 2A = 2\cos^2 A - 1$. $\cos 46° = \cos 2(23°) = 2\cos^2 23° - 1 = 2(\cos 23°)^2 - 1 = 2z^2 - 1$.

Example 3

If sin $x = A$, find cos $2x$ in terms of A.

Using the identity $\cos 2x = 1 - \sin^2 x$, you get $\cos 2x = 1 - A^2$.

You may be expected to solve trigonometric equations on the Math Level 2 Subject Test by using your graphing calculator and getting answers that are decimal approximations. To solve any equation, enter each side of the equation into a function (Y_n), graph both functions, and find the point(s) of intersection on the indicated domain by choosing an appropriate window.

Example 4

Solve 2 in x + cos $2x$ = 2 sin^2 x – 1 for $0 \le x \le 2\pi$.

Enter $2\sin x + \cos 2x$ into Y_1 and $2\sin^2 x - 1$ into Y_2. Set Xmin $= 0$, Xmax $= 2\pi$, Ymin $= -4$, and Ymax $= 4$. Solutions (x-coordinates of intersection points) are 1.57, 3.67, and 5.76

Example 5

Find values of x on the interval [0,π] for which cos x < sin $2x$.

Enter each side of the inequality into a function, graph both, and find the values of x where the graph of $\cos x$ lies beneath the graph of $\sin 2x$: $0.52 < x < 1.57$ or $x > 2.62$.

EXERCISES

1. If $\sin x = \dfrac{2}{3}$ and $\cos x = -\dfrac{5}{9}$, find the value of $\sin 2x$.

(A) $-\dfrac{20}{27}$

(B) $-\dfrac{10}{27}$

(C) $\dfrac{10}{27}$

(D) $\dfrac{20}{27}$

(E) $\dfrac{4}{3}$

2. If $\tan A = \cot B$, then

(A) $A = B$
(B) $A = 90° + B$
(C) $B = 90° + A$
(D) $A + B = 90°$
(E) $A + B = 180°$

3. If $\cos x = \dfrac{\sqrt{3}}{2}$, find $\cos 2x$.

(A) -0.87
(B) -0.25
(C) 0
(D) 0.5
(E) 0.75

4. If $\sin 37° = z$, express $\sin 74°$ in terms of z.

(A) $2z\sqrt{1 - z^2}$

(B) $2z^2 + 1$
(C) $2z$
(D) $2z^2 - 1$

(E) $\dfrac{z}{\sqrt{1 - z^2}}$

5. If $\sin x = -0.6427$, what is $\csc x$?

(A) -1.64
(B) -1.56
(C) 0.64
(D) 1.56
(E) 1.70

6. For what value(s) of x, $0 < x < \dfrac{\pi}{2}$, is $\sin x < \cos x$?

(A) $x < 0.79$
(B) $x < 0.52$
(C) $0.52 < x < 0.79$
(D) $x > 0.52$
(E) $x > 0.79$

7. What is the range of the function $f(x) = 5 - 6\sin(\pi x + 1)$.

(A) $[-6,6]$
(B) $[-5,5]$
(C) $[-1,1]$
(D) $[-1,11]$
(E) $[-11,1]$

INVERSE TRIG FUNCTIONS

If the graph of any trigonometric function $f(x)$ is reflected about the line $y = x$ (see Section 1.3), the graph of the inverse of that trigonometric function, $f^{-1}(x)$, results. These inverse trigonometric functions are called $\sin^{-1}$ or arcsin, $\cos^{-1}$ or arccos, and so on. In every case the resulting graph is *not* the graph of a function. To obtain a function, the range of the inverse relation must be severely limited. The particular range of each inverse trigonometric function is accepted by convention. The ranges of the inverse functions are as follows:

$$-\frac{\pi}{2} \le \sin^{-1} x \le \frac{\pi}{2}$$

$$0 \le \cos^{-1} x \le \pi$$

$$-\frac{\pi}{2} < \tan^{-1} x < \frac{\pi}{2}$$

$$0 < \cot^{-1} x < \pi$$

$$0 \le \sec^{-1} x \le \pi \text{ and } \sec^{-1} x \ne \frac{\pi}{2}$$

$$-\frac{\pi}{2} \le \csc^{-1} x \le \frac{\pi}{2} \text{ and } \csc^{-1} x \ne 0$$

The last three inverse functions are rarely used.

The inverse trig functions are used to represent angles with known trig values. If you know that the tangent of an angle is $\frac{8}{9}$, but you do not know the degree measure or radian measure of the angle, $\tan^{-1} \frac{8}{9}$ is an expression that represents an angle between $\frac{-\pi}{2}$ and $\frac{\pi}{2}$ whose tangent is $\frac{8}{9}$.

You can use your graphing calculator to find the degree or radian measure of an inverse trig value.

Example 1

Evaluate the radian measure of $\tan^{-1} \dfrac{8}{9}$.

Enter 2nd tan $\left(\dfrac{8}{9}\right)$ with you calculator in radian mode to get 0.73 radians.

Example 2

Evaluate the degree measure of $\sin^{-1} 0.8759$.

Enter 2nd sin (.8759) with your calculator in degree mode to get 61.15°.

Example 3

Evaluate the degree measure of $\sec^{-1} 3.4735$.

First define $x = \sec^{-1} 3.4735$. If $\sec x = 3.4735$, then $\cos x = \dfrac{1}{3.4735}$. Therefore, enter 2nd

$\cos\left(\dfrac{1}{3.4735}\right)$ with your calculator in degree mode to get $73.27°$.

If "trig" is any trigonometric function, $\text{trig}(\text{trig}^{-1}x) = x$. However, because of the range restriction on inverse trig functions, $\text{trig}^{-1}(\text{trig }x)$ need *not* equal x.

Example 4

Evaluate cos (cos⁻¹ 0.72).

$\cos(\cos^{-1} 0.72) = 0.72$.

Example 5

Evaluate sin⁻¹ (sin 265°).

Enter 2nd $\sin^{-1}(\sin(265))$ with your calculator in degree mode to get $-85°$. This is because $-85°$ is in the required range $[-90°, 90°]$, and $-85°$ has the same reference angle as $265°$.

Example 6

Evaluate $\sin\left(\cos^{-1}\dfrac{3}{5}\right)$.

Let $x = \cos^{-1}\dfrac{3}{5}$. Then $\cos x = \dfrac{3}{5}$ and x is in the first quadrant. See the figure below.

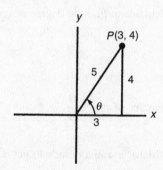

Use the Pythagorean identity $\sin^2 x + \cos^2 x = 1$ and the fact that x is in the first quadrant to

get $\sin x = \sqrt{1 - \cos^2 x} = \sqrt{\dfrac{16}{25}} = \dfrac{4}{5}$.

EXERCISES

1. Find the number of degrees in $\sin^{-1}\dfrac{\sqrt{2}}{2}$.

    (A) −45
    (B) −22.5
    (C) 0
    (D) 22.5
    (E) 45

2. Find the number or radians in $\cos^{-1}(-0.5624)$.

    (A) −0.97
    (B) 0.97
    (C) 1.77
    (D) 2.16
    (E) none of these

3. Evaluate $\tan^{-1}(\tan 128°)$.

    (A) −128°
    (B) −52°
    (C) 52°
    (D) 128°
    (E) none of these

4. Find the number of radians in $\cot^{-1}(-5.2418)$

    (A) −10.80
    (B) −5.30
    (C) −1.38
    (D) −0.19
    (E) none of these

5. Which of the following is (are) true?

    I. $\sin^{-1}1 + \sin^{-1}(-1) = 0$
    II. $\cos^{-1}1 + \cos^{-1}(-1) = 0$
    III. $\cos^{-1}x = \cos^{-1}(-x)$ for all x in the domain of $\cos^{-1}$

    (A) only I
    (B) only II
    (C) only III
    (D) only I and II
    (E) only II and III

6. Which of the following is a solution of $\cos 3x = \frac{1}{2}$?

(A) $60°$

(B) $\frac{5\pi}{3}$

(C) $\cos^{-1}\left(\frac{1}{6}\right)$

(D) $\cos^{-1}\left(\frac{\sqrt{3}}{2}\right)$

(E) $\frac{1}{3}\cos^{-1}\left(\frac{1}{2}\right)$

TRIANGLES

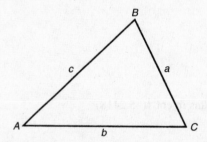

The final topic in trigonometry concerns the relationship between the angles and sides of a triangle that is *not* a right triangle. Depending on which of the sides and angles of the triangle are supplied, the following formulas can be used to find missing parts of a triangle. In $\triangle ABC$

$$\text{Law of Sines: } \frac{\sin A}{a} = \frac{\sin B}{b} = \frac{\sin C}{c}$$

used when the lengths of two sides and the value of the angle opposite one, or two angles and the length of one side are given.

$$\text{Law of Cosines: } a^2 = b^2 + c^2 - 2bc \cos A$$

$$b^2 = a^2 + c^2 - 2ac \cos B$$

$$c^2 = a^2 + b^2 - 2ab \cos C$$

used when the lengths of two sides and the included angle, or the lengths of three sides, are given.

$$\text{Area of a } \triangle: \text{Area} = \frac{1}{2} bc \sin A$$

$$\text{Area} = \frac{1}{2} ac \sin B$$

$$\text{Area} = \frac{1}{2} ab \sin C$$

used when two sides and the included angle are given.

Example 1

Find the number of degrees in the largest angle of a triangle whose sides are 3, 5, and 7.

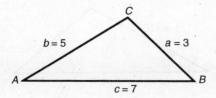

The largest angle is opposite the longest side. Use the Law of Cosines:

$$c^2 = a^2 + b^2 - 2ab \cdot \cos C$$
$$49 = 9 + 25 - 30 \cdot \cos C$$

Therefore, $\cos C = -\dfrac{15}{30} = -\dfrac{1}{2}$.

Since $\cos C < 0$ and $\angle C$ is an angle of a triangle, $90° < \angle C < 180°$.
Therefore, $\angle C = 120°$.

Example 2

Find the number of degrees in the other two angles of $\triangle ABC$ if $c = 75\sqrt{2}$, $b = 150$, and $\angle C = 30°$.

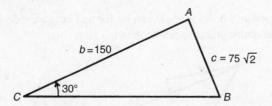

Use the law of sines:

$$\frac{75\sqrt{2}}{\sin 30°} = \frac{150}{\sin B}$$

$$75\sqrt{2} \cdot \sin B = 150 \cdot \sin 30°$$

$$\sin B = \frac{150 \cdot \dfrac{1}{2}}{75\sqrt{2}} = \frac{75}{75\sqrt{2}} = \frac{1}{\sqrt{2}} = \frac{\sqrt{2}}{2}$$

Therefore, $\angle B = 45°$ or $135°$; $\angle A = 105°$ or $15°$ since there are $180°$ in the sum of the three angles of a triangle.

Example 3

Find the area of $\triangle ABC$ if $a = 180$ inches, $b = 150$ inches, and $\angle C = 30°$.

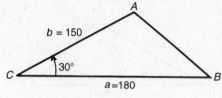

$$\text{Area} = \frac{1}{2}ab \cdot \sin C$$

$$= \frac{1}{2} \cdot 180 \cdot 150 \cdot \sin 30°$$

$$= \frac{1}{2} \cdot 180 \cdot 150 \cdot \frac{1}{2}$$

$$= 90 \cdot 75 = 675 \text{ square inches.}$$

Ambiguous Cases

If the lengths of two sides of a triangle and the angle opposite one of those sides are given, it is possible that two triangles, one triangle, or no triangle can be constructed with the data. This is called the *ambiguous* case. If the lengths of sides a and b and the value of $\angle A$ are given, the length of side b determines the number of triangles that can be constructed.

Case 1: If $\angle A > 90°$ and $a \le b$, no triangle can be formed because side a would not reach the base line. If $a > b$, one obtuse triangle can be drawn.

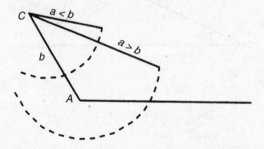

Let the length of the altitude from C to the base line be h. From the basic definition of sine, $\sin A = \dfrac{h}{b}$, and thus, $h = b \sin A$.

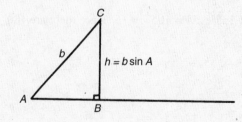

Case 2: If $\angle A < 90°$ and side $a < b \sin A$, no triangle can be formed. If $a = b \sin A$, one triangle can be formed. If $a > b$, there also will be only one triangle. If, on the other hand, $b \sin A < a < b$, two triangles can be formed.

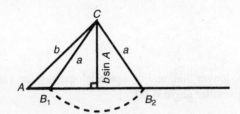

If a compass is opened the length of side a and a circle is drawn with center at C, the circle will cut the baseline at two points, B_1 and B_2. Thus, $\triangle AB_1C$ satisfies the conditions of the problem, as does $\triangle AB_2C$.

Example 1

How many triangles can be formed if $a = 24$, $b = 31$, and $\angle A = 30°$?

Because $\angle A < 90°$, $b \cdot \sin A = 31 \cdot \sin 30° = 31 \cdot \dfrac{1}{2} = 15\dfrac{1}{2}$. Since $b \cdot \sin A < a < b$, there are two triangles.

Example 2

How many triangles can be formed if $a = 24$, $b = 32$, and $\angle A = 150°$?

Since $\angle A > 90°$ and $a < b$, no triangle can be formed.

EXERCISES

1. In $\triangle ABC$, $\angle A = 30°$, $b = 8$, and $a = 4\sqrt{2}$. Angle C could equal

    (A) 45°
    (B) 135°
    (C) 60°
    (D) 15°
    (E) 90°

2. In $\triangle ABC$, $\angle A = 30°$, $a = 6$, and $c = 8$. Which of the following must be true?

    (A) $0° < \angle C < 90°$
    (B) $90° < \angle C < 180°$
    (C) $45° < \angle C < 135°$
    (D) $0° < \angle C < 45°$ or $90° < \angle C < 135°$
    (E) $0° < \angle C < 45°$ or $130° < \angle C < 180°$

3. The angles of a triangle are in a ratio of $8 : 3 : 1$. The ratio of the longest side of the triangle to the next longest side is

    (A) $\sqrt{6} : 2$
    (B) $8 : 3$
    (C) $\sqrt{3} : 1$
    (D) $8 : 5$
    (E) $2\sqrt{2} : \sqrt{3}$

4. The sides of a triangle are in a ratio of 4 : 5 : 6. The smallest angle is

    (A) 82°
    (B) 69°
    (C) 56°
    (D) 41°
    (E) 27°

5. Find the length of the longer diagonal of a parallelogram if the sides are 6 inches and 8 inches and the smaller angle is 60°.

    (A) 8
    (B) 11
    (C) 12
    (D) 7
    (E) 17

6. What are all values of side *a* in the figure below such that two triangles can be constructed?

    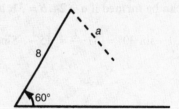

    (A) $a > 4\sqrt{3}$

    (B) $a > 8$

    (C) $a = 4\sqrt{3}$

    (D) $4\sqrt{3} < a < 8$

    (E) $8 < a < 8\sqrt{3}$

7. In $\triangle ABC$, $\angle B = 30°$, $\angle C = 105°$, and $b = 10$. The length of side *a* equals

    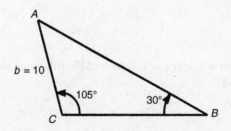

    (A) 7
    (B) 9
    (C) 10
    (D) 14
    (E) 17

8. The area of $\triangle ABC$, $= 24\sqrt{3}$, side $a = 6$, and side $b = 16$. The value of $\angle C$ is

   (A) 30°
   (B) 30° or 150°
   (C) 60°
   (D) 60° or 120°
   (E) none of the above

9. The area of $\triangle ABC = 12\sqrt{3}$, side $a = 6$, and side $b = 8$. Side $c =$

   (A) $2\sqrt{37}$

   (B) $2\sqrt{13}$

   (C) $2\sqrt{37}$ or $2\sqrt{13}$

   (D) 10

   (E) 10 or 12

10. Given the following data, which can form two triangles?

    I. $\angle C = 30°$, $c = 8$, $b = 12$

    II. $\angle B = 45°$, $a = 12\sqrt{2}$, $b = 15\sqrt{2}$

    III. $\angle C = 60°$, $b = 12$, $c = 5\sqrt{3}$

    (A) only I
    (B) only II
    (C) only III
    (D) only I and II
    (E) only I and III

Answers and Explanations

In these solutions the following notation is used:

   i: calculator unnecessary
   a: calculator helpful or necessary
   g: graphing calculator helpful or necessary

Definitions

1. i   **(A)** Reference angle is 40°. Cosine in quadrant IV is positive.

2. i   **(D)** See corresponding figure. Therefore, $\sin \theta = \dfrac{12}{13}$.

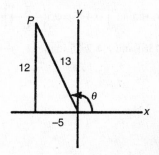

3. i   **(D)** Angle θ is in quadrant II since sec < 0 and sin > 0. Therefore, $\tan \theta = \dfrac{3}{-4} = -\dfrac{3}{4}$.

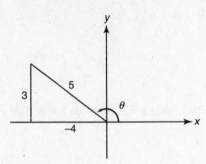

4. i   **(A)** Cofunctions of complementary angles are equal. $x - 30 + x = 90$ finds a reference angle of $60°$ for x. The angle in quadrant III that has a reference angle of $60°$ is $240°$.

5. i   **(A)** Angle α is in quadrant II, and sin α is positive. Angle β is in quadrant IV, and sin β is negative.

6. a   **(E)** Put your calculator in degree mode, $\cos 310° + \cos 190° \approx 0.643 + (-0.985) \approx -0.342$. Checking the answer choices shows that $-\cos 70° \approx -0.342$.

Arcs and Angles

1. i   **(E)** $30\left(\dfrac{180°}{\pi}\right) = \dfrac{5400°}{\pi}$.

2. i   **(D)** $s = r\theta$. $2\pi = r\theta$. $A = \dfrac{1}{2}r^2\theta$.

$$6\pi = \dfrac{1}{2}r^2\theta = \dfrac{1}{2}r(r\theta) = \dfrac{1}{2}r(2\pi). \qquad r = 6.$$

3. a   **(C)** $C = 2\pi r = 16.\ r = \dfrac{8}{\pi} \approx 2.55$.

$$A = \dfrac{1}{2}r^2\theta \approx \dfrac{1}{2}(2.55)^2(4.7) \approx 15.$$

4. a   **(A)** $40° = \dfrac{2\pi}{9} \approx 0.7$
$$s = r\theta \approx 1(0.7) \approx 0.7.$$

5. a   **(C)** Change $25°$ to 0.436 radian $\left(0.436 = 25\left(\dfrac{\pi}{180}\right)\right)$.

$s = r\theta$, and so $12 = r(0.436)$ and $r = 27.5$ inches.

Special Angles

1. i **(A)** Sketch a –60° angle in standard position as shown in the figure below.

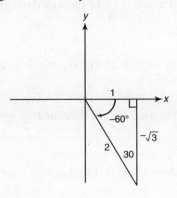

The tangent ratio is $\dfrac{y}{x} = \dfrac{-\sqrt{3}}{1} = -\sqrt{3}$.

2. i **(C)** Sketch an angle of $\dfrac{3\pi}{4}$ radians in standard position, as shown in the figure below.

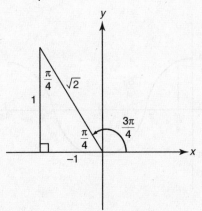

The cosine ratio is $\dfrac{x}{r} = \dfrac{-1}{\sqrt{2}} = \dfrac{-\sqrt{2}}{2}$.

3. i **(E)** First, determine an angle between 0° and 360° that is coterminal with 540° by subtracting 360° from 540° repeatedly until the result is in this interval. In this case, one subtraction suffices. Since coterminal angles have the same trig values, csc 540° = csc 180°. Sketch the figure below

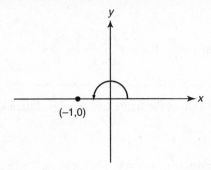

In a quadrantal angle $r = 1$, and the cosecant ratio is $\dfrac{r}{y}$, which is undefined.

Graphs

1. i   **(C)** Period = $\dfrac{2\pi}{2} = \pi$. Point P is $\dfrac{1}{4}$ of the way through the period. Amplitude is 1

because the coefficient of sin is 1. Therefore, point P is at $\left(\dfrac{\pi}{4}, 1\right)$.

2. i   **(E)** Amplitude = $\dfrac{1}{2}$. Period = π. Graph translated $\dfrac{1}{2}$ unit up. Graph looks like a

cosine graph reflected about x-axis and shifted up $\dfrac{1}{2}$ unit.

3. g   **(C)** Graph $4\cos\left(\dfrac{1}{2}x\right)$ using ZOOM/ZTRIG and observe that the portion of the graph

between $\dfrac{\pi}{4}$ and $\dfrac{5\pi}{4}$ is decreasing.

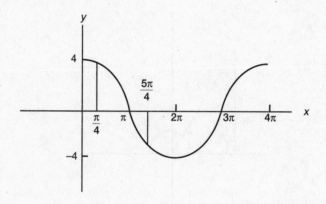

4. a   **(C)** Graph the function and determine its maximum (2) and minimum (–2). Subtract
and then divide by 2.

5. i   **(D)** Period = $\dfrac{2\pi}{P} = \dfrac{2\pi}{3}$.

6. g   **(C)** Graph the function using 0 for Xmin and $\dfrac{\pi}{2}$ for Xmax. Observe that the

maximum occurs when $x = \dfrac{\pi}{2}$. Then $f\left(\dfrac{\pi}{2}\right) = \sin\dfrac{\pi}{6} = \dfrac{1}{2}$.

7. i   **(D)** Period = $\dfrac{2\pi}{M} = 4\pi$ (from the figure), so $M = \dfrac{1}{2}$. Phase shift for a sine curve in

the figure is $-\pi$. Therefore, $\dfrac{1}{2}x + N = 0$ when $x = -\pi$. Therefore, $N = \dfrac{\pi}{2}$.

Identities, Equations, and Inequalities

1. a **(A)** $\sin 2x = 2\sin x \cos x = 2\left(\dfrac{2}{3}\right)\left(-\dfrac{5}{9}\right) = -\dfrac{20}{27}$.

2. i **(D)** Since tangent and cotangent are cofunctions, $\tan A = \cot(90° - A)$, so $B = 90° - A$, and $A + B = 90°$.

3. a **(D)** $\cos 2x = 2\cos^2 x - 1 = 2\left(\dfrac{\sqrt{3}}{2}\right)^2 - 1 = \dfrac{1}{2}$.

4. i **(A)** $\sin 74° = 2\sin 37° \cos 37°$. Since $\sin^2 x + \cos^2 x = 1$, $\cos x = \pm\sqrt{1 - \sin^2 x}$. Since

    $74°$ is in the first quadrant, the positive square root applies, so $\cos x = \sqrt{1 - z^2}$.

5. a **(B)** $\csc x = \dfrac{1}{\sin x} = \dfrac{1}{-0.6427} = -1.56$.

6. g **(A)** Graph $y = \sin x$ and $y = \cos$ in radian mode using the Xmin $= 0$ and Xmax $= \dfrac{\pi}{2}$.

    Observe that the first graph is beneath the second on $[0, 0.79]$.

7. i **(D)** Remember that the range of the sine function is $[-1,1]$, so the second term ranges from 6 to –6.

Inverse Trig Functions

1. a **(E)** Set your calculator to degree mode, and enter 2nd $\sin^{-1}\left(\sqrt{2}/2\right)$.

2. a **(D)** Set your calculator to radian mode, and enter 2nd $\cos^{-1}(-0.5624)$.

3. a **(B)** Set your calculator to degree mode, and enter 2nd $\tan^{-1}(\tan 128°)$.

4. a **(D)** Set your calculator to radian mode, and enter 2nd $\tan^{-1}\left(\dfrac{1}{-5.2418}\right)$.

5. i **(A)** Since $\sin^{-1}1 = \dfrac{\pi}{2}$ and $\sin^{-1}(-1) = -\dfrac{\pi}{2}$, I is true. Since $\cos^{-1}1 = 0$ and

    $\cos^{-1}(-1) = \pi$, II is not true. Since the range of $\cos^{-1}$ is $[0,\pi]$, III is not true because $\cos^{-1}$ can never be negative.

6. i **(E)** $3x = \arccos\left(\dfrac{1}{2}\right)$, and so $x = \dfrac{1}{3}\arccos\left(\dfrac{1}{2}\right)$.

Triangles

1. i   **(D)** Law of Sines: $\dfrac{\sin B}{8} = \dfrac{\frac{1}{2}}{4\sqrt{2}}$. $\sin B = \dfrac{\sqrt{2}}{2}$. $B = 45°$ or $135°$. The figure shows two

possible locations for B, labeled B_1 and B_2, where $m\angle AB_1C = 45°$ and $m\angle AB_2C = 135°$. Corresponding to these, $m\angle ACB_1 = 105°$ and $m\angle ACB_2 = 15°$. Of these, only $15°$ is an answer choice.

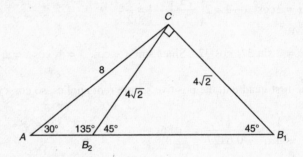

2. i   **(E)** Law of Sines: $\dfrac{\sin C}{8} = \dfrac{\frac{1}{2}}{6}$. $\sin C = \dfrac{2}{3} \approx 0.67$. $\sin 45° = \dfrac{\sqrt{2}}{2} \approx 0.7$. Since sine is

an increasing function in quadrant I, $0° \le C \le 45°$. Angles in quadrant II greater than $135°$ use these values of C as reference angles.

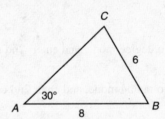

3. i   **(A)** The angles are $15°$, $45°$, and $120°$. Let c be the longest side and b the next

longest. $\dfrac{\sin 120°}{c} = \dfrac{\sin 45°}{b}$. $\dfrac{c}{b} = \dfrac{\sin 120°}{\sin 45°} = \dfrac{\frac{\sqrt{3}}{2}}{\frac{\sqrt{2}}{2}} = \dfrac{\sqrt{6}}{2}$.

4. a   **(D)** Use the Law of Cosines. Let the sides be 4, 5, and 6. $16 = 25 + 36 - 60 \cos A$.

Cos $A = \dfrac{45}{60} = \dfrac{3}{4}$, which implies that $A = \cos^{-1}(0.75) \approx 41°$.

5. a   **(C)** Law of Cosines: $d^2 = 36 + 64 - 96 \cos 120°$. $d^2 = 148$. Therefore, $d \approx 12$.

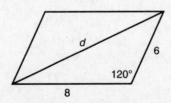

6. i **(D)** Altitude to base = $8 \sin 60° = 4\sqrt{3}$. Therefore, $4\sqrt{3} < a < 8$.

7. a **(D)** $A = 45°$. Law of Sines: $\dfrac{\sin 45°}{a} = \dfrac{\sin 30°}{10}$. Therefore, $a = 10\sqrt{2} \approx 14$.

8. i **(D)** Area $= \dfrac{1}{2}ab \sin C$. $24\sqrt{3} = \dfrac{1}{2} \cdot 6 \cdot 16 \sin C$. $\operatorname{Sin} C = \dfrac{\sqrt{3}}{2}$. Therefore, $C = 60°$ or $120°$.

9. i **(C)** Area $= \dfrac{1}{2}ab \sin C$. $12\sqrt{3} = \dfrac{1}{2} \cdot 6 \cdot 8 \sin C$. $\operatorname{Sin} C = \dfrac{\sqrt{3}}{2}$. $C = 60°$ or $120°$.

Use Law of Cosines with 60° and then with 120°.

Note: At this point in the solution you know there have to be two values for C. Therefore, the answer must be Choice C or E. If $C = 10$ (from Choice E), ABC is a right triangle with area $= \dfrac{1}{2} \cdot 6 \cdot 8 = 24$. Therefore, Choice E is not the answer, and so Choice C is the correct answer.

10. i **(A)** In I the altitude $= 12 \cdot \dfrac{1}{2} = 6$, $6 < c < 12$, and so 2 triangles. In II $b > 12\sqrt{2}$, so only 1 triangle. In III the altitude $= 12 \cdot \dfrac{\sqrt{3}}{2} > 5\sqrt{3}$, so no triangle.

1.4 Exponential and Logarithmic Functions

The basic properties of exponents and logarithms and the fact that the exponential function and the logarithmic function are inverses lead to many interesting problems.

The basic exponential properties:
For all positive real numbers x and y, and all real numbers a and b:

$$x^a \cdot x^b = x^{a+b} \quad x^0 = 1$$
$$\frac{x^a}{x^b} = x^{a-b} \quad x^{-a} = \frac{1}{x^a}$$
$$\left(x^a\right)^b = x^{ab} \quad x^a \cdot y^a = (xy)^a$$

The basic logarithmic properties:
For all positive real numbers a, b, p, and q, and all real numbers x, where $a \neq 1$ and $b \neq 1$:

$$\log_b(p \cdot q) = \log_b p + \log_b q \quad \log_b 1 = 0 \quad b^{\log_b p} = p$$
$$\log_b\left(\frac{p}{q}\right) = \log_b p - \log_b q \quad \log_b b = 1$$
$$\log_b\left(p^x\right) = x \cdot \log_b p \quad \log_b p = \frac{\log_a p}{\log_a b}$$

NOTE

$\log_b x$ is only defined for positive x

The basic property that relates the exponential and logarithmic functions is:

For all real numbers x, and all positive real numbers b and N, $\log_b N = x$ is equivalent to $b^x = N$.

By convention, the base is 10 if no base is indicated.

Example 1

Simplify $x^{n-1} \cdot x^{2n} \cdot \left(x^{2-n}\right)^2$

This is equal to $x^{n-1} \cdot x^{2n} \cdot x^{4-2n} = x^{n-1+2n+4-2n} = x^{n+3}$.

Example 2

Simplify $\dfrac{3^{n-2} \cdot 9^{2-n}}{3^{2-n}}$.

In order to combine exponents using the properties above, the base of each factor must be the same.

$$\frac{3^{n-2} \cdot 9^{2-n}}{3^{2-n}} = \frac{3^{n-2} \cdot (3^2)^{2-n}}{3^{2-n}} = \frac{3^{n-2} \cdot 3^{4-2n}}{3^{2-n}}$$
$$= 3^{n-2+4-2n-(2-n)} = 3^0 = 1$$

Note: Examples 3 and 4 can be easily evaluated with a calculator.

Example 3

If log 23 = z, what does log 2300 equal?

$\log 2300 = \log(23 \cdot 100) = \log 23 + \log 100 = z + \log 10^2 = z + 2$

Example 4

If $\log_b 2 = x$ and $\log_b 3 = y$, find the value of $\log_b 18$ in terms of x and y.

$\log_b 18 = \log_b(3^2 \cdot 2) = 2\log_b 3 + \log_b 2 = 2x + y$

Example 5

Solve for x: $\log_b(x + 5) = \log_b x + \log_b 5$.

$\log_b x + \log_b 5 = \log_b (5x)$

Therefore, $\log(x + 5) = \log(5x)$, which is true only when:

$$x + 5 = 5x$$
$$5 = 4x$$
$$x = \frac{5}{4}$$

Example 6

Evaluate $\log_{27} \sqrt{54} - \log_{27} \sqrt{6}$.

$$\log_{27} \sqrt{54} - \log_{27} \sqrt{6} = \log_{27}\left(\frac{\sqrt{54}}{\sqrt{6}}\right)$$

$$= \log_{27} \sqrt{9} = \log_{27} 3 = x$$

The last equality implies that

$$27^x = 3$$
$$(3^3)^x = 3$$
$$3^{3x} = 3^1$$

Therefore, $3x = 1$ and $x = \dfrac{1}{3}$.

Thus, $\log_{27} \sqrt{54} - \log_{27} \sqrt{6} = \dfrac{1}{3}$.

You could also use the change-of-base formula and your calculator.

$$\log_{27} \sqrt{54} = \frac{\log_{10} \sqrt{54}}{\log_{10} 27} \approx \frac{0.866}{1.431} \approx 0.605$$

$$\log_{27} \sqrt{6} = \frac{\log_{10} \sqrt{6}}{\log_{10} 27} \approx \frac{0.389}{1.431} \approx 0.272$$

Therefore, $\log_{27} \sqrt{54} - \log_{27} \sqrt{6} \approx 0.333 \approx \dfrac{1}{3}$.

The graphs of all exponential functions $y = b^x$ have roughly the same shape and pass through point $(0,1)$. If $b > 1$, the graph increases as x increases and approaches the x-axis as an asymptote as x decreases. The amount of curvature becomes greater as the value of b is made greater. If $0 < b < 1$, the graph increases as x decreases and approaches the x-axis as an asymptote as x increases. The amount of curvature becomes greater as the value of b is made closer to zero.

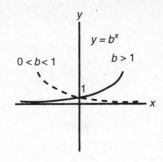

The graphs of all logarithmic functions $y = \log_b x$ have roughly the same shape and pass through point (1,0). If $b > 1$, the graph increases as x increases and approaches the y-axis as an asymptote as x approaches zero. The amount of curvature becomes greater as the value of b is made greater. If $0 < b < 1$, the graph decreases as x increases and approaches the y-axis as an asymptote as x approaches zero. The amount of curvature becomes greater as the value of b is made closer to zero.

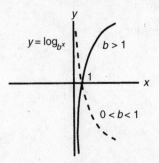

EXERCISES

1. If $x^a \cdot (x^{a+1})^a \cdot (x^a)^{1-a} = x^k$, then $k =$

   (A) $2a + 1$
   (B) $a + a^2$
   (C) $3a$
   (D) $3a + 1$
   (E) $a^3 + a$

2. If $\log_8 3 = x \cdot \log_2 3$, then $x =$

   (A) $\dfrac{1}{3}$

   (B) 3
   (C) 4
   (D) $\log_4 3$
   (E) $\log_8 9$

3. If $\log_{10} m = \dfrac{1}{2}$, then $\log_{10} 10m^2 =$

   (A) 2
   (B) 2.5
   (C) 3
   (D) 10.25
   (E) 100

4. If $\log_b 5 = a$, $\log_b 2.5 = c$, and $5^x = 2.5$, then $x =$

   (A) ac

   (B) $\dfrac{c}{a}$

   (C) $a + c$
   (D) $c - a$
   (E) The value of x cannot be determined from the information given.

5. If $f(x) = \log_2 x$, then $f\left(\dfrac{2}{x}\right) + f(x) =$

(A) $\log\left(\dfrac{2}{x}\right) + \log_2 x$

(B) 1

(C) $\log_2\left(\dfrac{2 + x^2}{x}\right)$

(D) $\log_2\left(\dfrac{2}{x}\right) \cdot \log_2 x$

(E) 0

6. If $\log_b (xy) < 0$, which of the following must be true?

(A) $xy < 0$
(B) $xy < 1$
(C) $xy > 1$
(D) $xy > 0$
(E) none of the above

7. If $\log_2 m = \sqrt{7}$ and $\log_7 n = \sqrt{2}$, $mn =$

(A) 1
(B) 2
(C) 96
(D) 98
(E) 103

8. $\text{Log}_7 5 =$

(A) 1.2
(B) 1.1
(C) 0.9
(D) 0.8
(E) -0.7

9. $\left(\sqrt[3]{2}\right)\left(\sqrt[5]{4}\right)\left(\sqrt[9]{8}\right) =$

(A) 1.9
(B) 2.0
(C) 2.1
(D) 2.3
(E) 2.5

10. If $300 is invested at 3%, compounded continuously, how long (to the nearest year) will it take for the money to double? (If P is the amount invested, the formula for the amount, A, that is available after t years is $A = Pe^{0.03t}$.)

(A) 26
(B) 25
(C) 24
(D) 23
(E) 22

Answers and Explanations

In these solutions the following notation is used:

i: calculator unnecessary
a: calculator helpful or necessary
g: graphing calculator helpful or necessary

Exponential and Logarithmic Functions

1. i **(C)** $x^a \cdot x^{a^2+a} \cdot x^{a-a^2} = x^{a+a^2+a+a-a^2} = x^{3a}$.

2. i **(A)** Let $\log_8 3 = y$. Then $\log_2 3^x = y$. Then $8^y = 3$ implies $2^{3y} = 3$, and $2^y = 3^{1/3}$. From the second equation $2^y = 3^x$. Therefore, $3^x = 3^{1/3}$, and so $x = \dfrac{1}{3}$.

3. i **(A)** $\log(10m^2) = \log 10 + 2 \log m = 1 + 2 \cdot \dfrac{1}{2} = 2.$

4. i **(B)** $b^a = 5$, $b^c = 2.5 = 5^x$, using the relationships between logs and exponents: $(b^a)^x = b^{ax} = 5^x = b^c$. Therefore, $ax = c$ and $x = \dfrac{c}{a}$.

5. i **(B)** $f\left(\dfrac{2}{x}\right) + f(x) = \log_2\left(\dfrac{2}{x}\right) + \log_2 x$
$$= \log_2 2 - \log_2 x + \log_2 x = 1.$$

6. i **(E)** If $b > 1$, Choice B is the answer. If $b < 1$, Choice C is the answer. Since no restriction was put on b, however, the correct answer is Choice E.

7. a **(D)** Converting the log expressions to exponential expressions gives $m = 2^{\sqrt{7}}$ and $n = 7^{\sqrt{2}}$. Therefore, $mn = 2^{\sqrt{7}} \cdot 7^{\sqrt{2}} \approx 6.2582 \cdot 15.673 \approx 98$.

8. a **(D)** $\log_7 5 = \dfrac{\log 5}{\log 7} \approx \dfrac{0.699}{0.845} \approx 0.8 \cdot$

9. a **(C)** $\sqrt[3]{2}\sqrt[5]{4}\sqrt[9]{8} = 2^{1/3}4^{1/5}8^{1/9} \approx 2.1.$

10. a **(D)** Substitute in $A = Pe^{0.03t}$ to get $600 = 300e^{0.03t}$. Simplify to get $2 = e^{0.03t}$. Then take ln of both sides to get $\ln 2 = 0.03t$ and $t = \dfrac{\ln 2}{0.03}$. Use your calculator to find that t is approximately 23.

1.5 Rational Functions and Limits

F is a rational function if and only if $F(x) = \dfrac{p(x)}{q(x)}$, where $p(x)$ and $q(x)$ are both polynomial

functions and $q(x)$ is not zero. As a general rule, the graphs of rational functions are not continuous (i.e., they have holes, or sections of the graphs are separated from other sections by asymptotes). A point of discontinuity occurs at any value of x that would cause $q(x)$ to become zero.

If $p(x)$ and $q(x)$ can be factored so that $F(x)$ can be reduced, removing the factors that caused the discontinuities, the graph will contain only holes. If the factors that caused the discontinuities cannot be removed, asymptotes will occur.

TIP

If p and q are both 0 for some x, there is a hole at that x, but if q is 0 and p is not 0, these is an asymptote.

Example 1

Sketch the graph of $F(x) = \dfrac{x^2 - 1}{x + 1}$ **.**

There is a discontinuity at $x = -1$ since this value would cause division by zero. The fraction
$\dfrac{x^2 - 1}{x + 1} = \dfrac{(x - 1)(x + 1)}{(x + 1)} = (x - 1)$, and so the graph of $F(x)$ is the same as the graph of
$y = x - 1$ except for a hole at $x = -1$.

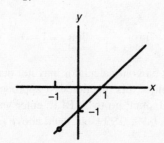

You could also enter this function on the graphing calculator and get the graph shown above, but it is unlikely that you would see the hole at $(-1, 0)$.

TIP

Don't count on seeing holes on a graphing calculator.

Example 2

Sketch the graph of $F(x) = \dfrac{1}{x - 2}$ **.**

Since this fraction cannot be reduced, and $x = 2$ would cause division by zero, a vertical asymptote occurs when $x = 2$. This is true because, as x approaches very close to 2, $F(x)$ gets either extremely large or extremely small. As x becomes extremely large or extremely small, $f(x)$ gets closer and closer to zero. This means that a horizontal asymptote occurs when $y = 0$. Plotting a few points indicates that the graph looks like the figure below.

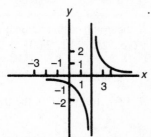

Vertical and horizontal asymptotes can also be described using limit notation. In this example you could write

$$\lim_{x \to +\infty} f(x) = 0 \text{ to mean } f(x) \text{ gets closer to 0 as } x \text{ gets arbitrarily large or small}$$

$$\lim_{x \to 2^+} f(x) = \infty \text{ to mean } f(x) \text{ gets arbitrarily large as } x \text{ approaches 2 from the right}$$

$$\lim_{x \to 2^-} f(x) = -\infty \text{ to mean } f(x) \text{ gets arbitrarily small as } x \text{ approaches 2 from the left}$$

If you entered this function into a TI-83 calculator, your graph will show what appears to be an asymptote. (Actually, the graphing calculator connects the pixel of the largest *x*-coordinate for which the *y*-coordinate is negative to the pixel of the smallest *x*-coordinate for which the *y*-pixel is positive.) The TI-84 calculators do not connect these two pixels. In either case, you can "read" the graph to determine the two infinite limits.

Example 3

What does $\displaystyle\lim_{x \to 1} \frac{x^2 - 1}{x + 1}$ equal?

Since $\dfrac{x^2 - 1}{x + 1}$ reduces to $x - 1$,

$$\lim_{x \to 1} \frac{x^2 - 1}{x + 1} = \lim_{x \to 1} x - 1 = 0.$$

Although the window settings on your calculator may not make it possible to see the hole at *x* = 1, you can determine the limit of this function as *x* approaches 1 from both sides by using the table feature. Select Ask for Indpnt; go to TABLE; enter values of *x* that get progressively closer to 1 from below (e.g., 0.9, 0.99, 0.999, etc.) and above (e.g., 1.1, 1.01, 1.001, etc.); and watch *y* get closer to 1.

Example 4

What does $\displaystyle\lim_{x \to 2^+} 3x + 5$ equal?

Since "problems" occur only when division by zero appears imminent, this example is extremely easy. As *x* gets closer and closer to 2, 3*x* + 5 seems to be approaching closer and closer to 11. Therefore, $\displaystyle\lim_{x \to 2^+} 3x + 5 = 11$.

Example 5

What does $\displaystyle\lim_{x \to 2} \left(\frac{3x^2 + 5}{x - 2} \right)$ equal?

The numerator is always positive, so the graph of this rational function has a vertical asymptote. As *x* approaches 2 from the right (i.e., 2.1, 2.01, 2.001, . . .), the denominator approaches zero from the right so $\dfrac{3x^2 + 5}{x - 2}$ gets larger and larger and approaches positive infinity. As *x* approaches 2 from the left (i.e., 1.9, 1.99, 1.999, . . .), the denominator approaches zero from the left so $\dfrac{3x^2+5}{x-2}$ gets smaller and smaller and approaches negative infinity. Thus,

$\displaystyle\lim_{x\to 2^-}\dfrac{3x^2+5}{x-2}$ does not exist since $\displaystyle\lim_{x\to 2^-}\dfrac{3x^2+5}{x-2}=-\infty$ and $\displaystyle\lim_{x\to 2^+}\dfrac{3x^2+5}{x-2}=+\infty$.

As in the previous example, you could enter this function into your graphing calculator. Then, with an appropriate window, you could determine the infinite limits as x approaches 2 from the left and right.

Example 6

If $f(x)=\left\{\begin{array}{ll}3x+2 & \text{when } x\neq 0\\ 0 & \text{when } x=0\end{array}\right\}$, what does $\displaystyle\lim_{x\to 0} f(x)$ equal?

As x approaches zero, $3x+2$ approaches 2, in spite of the fact that $f(x)=0$ when $x=0$. Therefore, $\displaystyle\lim_{x\to 0} f(x)=2$.

Example 7

What does $\displaystyle\lim_{x\to\infty}\left(\dfrac{3x^2+4x+2}{2x^2+x-5}\right)$ equal?

As x gets larger and larger, the x^2 terms in the numerator and denominator "dominate" in the sense that the terms of lower degree become negligible. Therefore, the larger x gets, the more the rational function looks like $\dfrac{3x^2}{2x^2}=\dfrac{3}{2}$.

You can also see this by dividing each term of the numerator and denominator by x^2, the highest power of x.

$$\lim_{x\to\infty}\left(\dfrac{3x^2+4x+2}{2x^2+x-5}\right)=\lim_{x\to\infty}\left(\dfrac{3+\dfrac{4}{x}+\dfrac{2}{x^2}}{2+\dfrac{1}{x}-\dfrac{5}{x^2}}\right).$$

Now, as $x\to\infty, \dfrac{4}{x},\dfrac{2}{x^2},\dfrac{1}{x}$, and $\dfrac{5}{x^2}$, each approaches zero. Thus, the entire fraction approaches

$\dfrac{3+0+0}{2+0-0}=\dfrac{3}{2}$. Therefore, $\displaystyle\lim_{x\to\infty}\left(\dfrac{3x^2+4x+2}{2x^2+x-5}\right)=\dfrac{3}{2}$.

To use a graphing calculator to find this limit, enter the function, and use the TABLE in Ask mode to enter larger and larger x values. The table will show y values closer and closer to 1.5.

1. To be continuous at $x = 1$, the value of $\dfrac{x^4 - 1}{x^3 - 1}$ must be defined to be equal to

   (A) -1
   (B) 0
   (C) 1
   (D) $\dfrac{4}{3}$
   (E) 4

2. If $f(x) = \begin{cases} \dfrac{3x^2 + 2x}{x} & \text{when } x \neq 0 \\ k & \text{when } x = 0 \end{cases}$, what must the value of k be equal to in order for $f(x)$ to

   be a continuous function?

   (A) $-\dfrac{3}{2}$

   (B) $-\dfrac{2}{3}$

   (C) 0
   (D) 2
   (E) No value of k can make $f(x)$ a continuous function.

3. $\displaystyle\lim_{x \to 2}\left(\dfrac{x^3 - 8}{x^4 - 16}\right) =$

   (A) 0

   (B) $\dfrac{3}{8}$

   (C) $\dfrac{1}{2}$

   (D) $\dfrac{4}{7}$

   (E) This expression is undefined.

4. $\displaystyle\lim_{x \to \infty}\left(\dfrac{5x^2 - 2}{3x^2 + 8}\right) =$

   (A) $-\dfrac{1}{4}$

   (B) 0

   (C) $\dfrac{3}{11}$

   (D) $\dfrac{5}{3}$

   (E) ∞

5. Which of the following is the equation of an asymptote of $y = \dfrac{3x^2 - 2x - 1}{9x^2 - 1}$?

   (A) $x = -\dfrac{1}{3}$

   (B) $x = 1$

   (C) $y = -\dfrac{1}{3}$

   (D) $y = \dfrac{1}{3}$

   (E) $y = 1$

Answers and Explanations

In these solutions the following notation is used:

   i: calculator unnecessary
   a: calculator helpful or necessary
   g: graphing calculator helpful or necessary

Rational Functions and Limits

All of these exercises can be completed with the aid of a graphing calculator as described in the example.

1. i **(D)** Factor and reduce: $\dfrac{(x-1)(x+1)\left(x^2+1\right)}{(x-1)\left(x^2+x+1\right)}$. Substitute 1 for x and the fraction

    equals $\dfrac{4}{3}$.

2. i **(D)** Factor and reduce the fraction, which becomes $3x + 2$. As x approaches zero, this approaches 2.

3. i **(B)** Factor and reduce $\dfrac{(x-2)\left(x^2+2x+4\right)}{(x-2)(x+2)\left(x^2+4\right)}$. Substitute 2 for x and the fraction

    equals $\dfrac{3}{8}$.

4. i **(D)** Divide numerator and denominator through by x^2. As $x \to \infty$, the fraction

    approaches $\dfrac{5}{3}$.

5. i **(D)** Factor and reduce $\dfrac{(3x+1)(x-1)}{(3x+1)(3x-1)}$. Therefore a vertical asymptote occurs when

    $3x - 1 = 0$ or $x = \dfrac{1}{3}$, but this is not an answer choice. As $x \to \infty$, $y \to \dfrac{1}{3}$. Therefore,

    $y = \dfrac{1}{3}$ is the correct answer choice.

1.6 Miscellaneous Functions

PARAMETRIC EQUATIONS

At times, it is convenient to express a relationship between x and y in terms of a third variable, usually denoted by a **parameter** t. For example, **parametric equations** $x = x(t)$, $y = y(t)$ can be used to locate a particle on the plane as a various time at t.

Example 1

Graph the parametric equations $\begin{cases} x = 3t + 4 \\ y = t - 5 \end{cases}$

Select MODE on your graphing calculator, and select PAR. Enter $3t + 4$ into X_{1T} and $t - 5$ into Y_{1T}. The standard window uses 0 for Tmin and 6.28... (2π) for Tmax along with the usual ranges for x and y. The choice of 0 for Tmin reflects the interpretation of t as "time." With the standard window, the graph looks like the figure below.

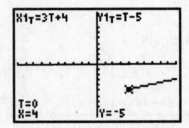

If you use TRACE, the cursor will begin at $t = 0$, where $(x,y) = (4,-5)$. As t increases from 0, the graph traces out a line that ascends as it moves right.

It may be possible to eliminate the parameter and to rewrite the equation in familiar xy-form. Just remember that the resulting equation may consist of points not on the graph of the original set of equations.

Example 2

$\begin{cases} x = t^2 \\ y = 3t^2 + 1 \end{cases}$ **Eliminate the parameter and sketch the graph.**

Substituting x for t^2 in the second equation results in $y = 3x + 1$, which is the equation of a line with a slope of 3 and a y-intercept of 1. However, the original parametric equations indicate that $x \geq 0$ and $y \geq 1$ since t^2 cannot be negative. Thus, the proper way to indicate this set of points without the parameter is as follows: $y = 3x + 1$ and $x \geq 0$. The graph is the ray indicated in the figure.

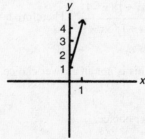

Example 3

Sketch the graph of the parametric equations $\begin{cases} x = 4\cos\theta \\ y = 3\sin\theta \end{cases}$

Replace the parameter θ with t, and enter the pair of equations. The graph has the shape of an ellipse, elongated horizontally, as shown in this diagram.

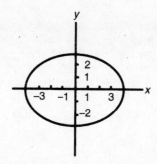

It is possible to eliminate the parameter, θ, by dividing the first equation by 4 and the second equation by 3, squaring each, and then adding the equations together.

$$\left(\frac{x}{4}\right)^2 = \cos^2\theta \quad \text{and} \quad \left(\frac{y}{3}\right)^2 = \sin^2\theta$$

$$\frac{x^2}{16} + \frac{y^2}{9} = \cos^2\theta + \sin^2\theta = 1$$

Here, $\frac{x^2}{16} + \frac{y^2}{9} = 1$ is the equation of an ellipse with its center at the origin, $a = 4$, and $b = 3$ (see Section 3.2). Since $-1 \le \cos\theta \le 1$ and $-1 \le \sin\theta \le 1$, $-4 \le x \le 4$ and $-3 \le y \le 3$ from the two parametric equations. In this case the parametric equations do not limit the graph obtained by removing the parameter.

EXERCISES

1. In the graph of the parametric equations $\begin{cases} x = t^2 + t \\ y = t^2 - t \end{cases}$

   (A) $x \ge 0$

   (B) $x \ge -\dfrac{1}{4}$

   (C) x is any real number
   (D) $x \ge -1$
   (E) $x \le 1$

2. The graph of $\begin{cases} x = \sin^2 t \\ y = 2\cos t \end{cases}$ is a

   (A) straight line
   (B) line segment
   (C) parabola
   (D) portion of a parabola
   (E) semicircle

3. Which of the following is (are) a pair of parametric equations that represent a circle?

   I. $\begin{cases} x = \sin \theta \\ y = \cos \theta \end{cases}$

   II. $\begin{cases} x = t \\ y = \sqrt{1 - t^2} \end{cases}$

   III. $\begin{cases} x = \sqrt{s} \\ y = \sqrt{1 - s} \end{cases}$

   (A) only I
   (B) only II
   (C) only III
   (D) only II and III
   (E) I, II, and III

PIECEWISE FUNCTIONS

Piecewise functions are defined by different equations on different parts of their domain. These functions are useful in modeling behavior that exhibits more than one pattern.

Example 1

Graph the function $f(x) = \begin{cases} 3 - x^2 \text{ if } x < 1 \\ x^3 - 4x \text{ if } x \geq 1 \end{cases}$

You can graph this on your graphing calculator by using the 2nd TEST command to enter the symbols $<$ and $\geq$. Enter $(3 - x^2)(x < 1) + (x^3 - 4x)(x \geq 1)$ into Y_1. For values of x less than 1, $(x - 1) = 1$ and $(x \geq 1) = 0$, so for these values only $3 - x^2$ will be graphed.

The reverse is true for values of x greater than 1, so only $x^3 - 4x$ will be graphed. This graph is shown on the standard grid in the figure below.

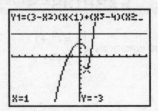

Absolute value functions are a special type of piecewise functions. The absolute value function is defined as $|x| = \begin{cases} x \text{ if } x \geq 0 \\ -x \text{ if } x < 0 \end{cases}$

The general absolute value function has the form $f(x) = a|x - h| + k$, with the vertex at (h,k) and shaped like $\vee$ if $a > 0$ and like $\wedge$ if $a < 0$. The vertex separates the two branches of the graph; h delineates the domain of all real numbers into two parts. The magnitude of a determines how spread out the two branches are. Larger values of a correspond to graphs that are more spread out.

The absolute value command is in the MATH/NUM menu of your graphing calculator. You can readily solve absolute value equations or inequalities by finding points of intersection.

Example 2

If |x – 3| = 2, find x.

Enter |x – 3| into Y_1 and 2 into Y_2. As seen in the figure below, the points of intersection are at $x = 5$ and $x = 1$.

This is also easy to see algebraically. If |x – 3| = 2, then $x - 3 = 2$ or $x - 3 = -2$. Solving these equations yields the same solutions: 5 or 1. This equation also has a coordinate geometry solution: |$a - b$| is the distance between a and b. Thus |x – 3| = 2 has the interpretation that x is 2 units from 3. Therefore, x must be 5 or 1.

Example 3

Find all values of x for which |$2x + 3$| ≥ 5.

The graphical solution is shown below.

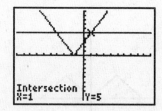

The desired values of x are on, or right and left of, the points of intersection: $x \geq 1$ or $x \leq -4$.

By writing the inequality as $\left| x - \left(-\dfrac{3}{2} \right) \right| \geq \dfrac{5}{2}$, we can also interpret the solutions to the inequality as those points that are more than $2\dfrac{1}{2}$ units from $-1\dfrac{1}{2}$.

Example 4

If the graph of $f(x)$ is shown below, sketch the graph of (A) |$f(x)$| (B) $f(|x|)$.

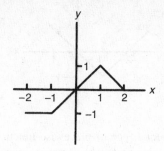

(A) Since $|f(x)| \geq 0$, by the definition of absolute value, the graph cannot have any points below the x-axis. If $f(x) < 0$, then $|f(x)| = -f(x)$. Thus, all points below the x-axis are reflected about the x-axis, and all points above the x-axis remain unchanged.

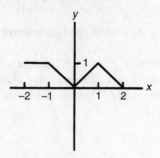

(B) Since the absolute value of x is taken before the function value is found, and since $|x| = -x$ when $x < 0$, any negative value of x will graph the same y-values as the corresponding positive values of x. Thus, the graph to the left of the y-axis will be a reflection of the graph to the right of the y-axis.

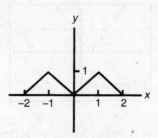

Example 5

If $f(x) = |x + 1| - 1$, what is the minimum value of $f(x)$?

Since $|x + 1| \geq 0$, its smallest value is 0. Therefore, the smallest value of $f(x)$ is $0 - 1 = -1$. The graph of $f(x)$ is indicated below.

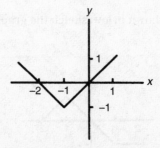

Step functions are another special type of piecewise function. These functions are constant over different parts of their domains so that their graphs consist of horizontal segments. The greatest integer function, denoted by $[x]$, is an example of a step function. If x is an integer, then $[x] = x$. If x is not an integer, then $[x]$ is the largest integer less than x.

The greatest integer function is in the MATH/NUM menu as int on TI-83/84 calculators.

Example 6

(A) [3.2] = 3
(B) [1.999] = 1
(C) [5] = 5
(D) [−3.12] = −4
(E) [−0.123] = − 1.

Example 7

Sketch the graph of $f(x) = [x]$.

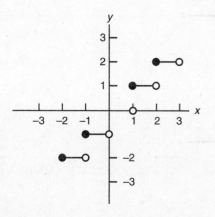

> **CAUTION**
>
> On the graphing calculator, you can't tell which side of each horizontal segment has the open and closed point.

Example 8

What is the range of $f(x) = \left[\dfrac{[x]}{x}\right]$.

Enter int(int(x)/x) into Y_1 and choose Auto for both Indpnt and Depend in TBLSET. Set TblStart to 0 and $\triangle$Tbl to 0.1. Inspection of TABLE shows only 0s and 1s as Y_1, so the range is the two-point set {0,1}.

EXERCISES

1. $|2x − 1| = 4x + 5$ has how many numbers in its solution set?

   (A) 0
   (B) 1
   (C) 2
   (D) an infinite number
   (E) none of the above

2. Which of the following is equivalent to $1 \le |x − 2| \le 4$?

   (A) $3 \le x \le 6$
   (B) $x \le 1$ or $x \ge 3$
   (C) $1 \le x \le 3$
   (D) $x \le −2$ or $x \ge 6$
   (E) $−2 \le x \le 1$ or $3 \le x \le 6$

3. The area bound by the relation $|x| + |y| = 2$ is

(A) 8
(B) 1
(C) 2
(D) 4
(E) There is no finite area.

4. Given a function, $f(x)$, such that $f(x) = f(|x|)$. Which one of the following could be the graph of $f(x)$?

(A)

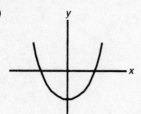

(B)

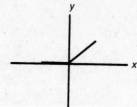

(C)

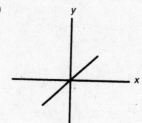

(D)

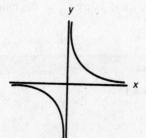

(E)

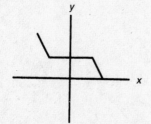

5. The figure shows the graph of which one of the following?

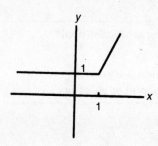

    (A) $y = 2x - |x|$
    (B) $y = |x - 1| + x$
    (C) $y = |2x - 1|$
    (D) $y = |x + 1| - x$
    (E) $y = 2|x| - |x|$

6. The postal rate for first-class mail is 41 cents for the first ounce or portion thereof and 17 cents for each additional ounce or portion thereof up to 3.5 ounces. The cost in cents of first-class postage for a letter weighing N ounces ($N \le 3.5$) is

    (A) $41 + [N - 1] \cdot 17$
    (B) $[N - 41] \cdot 17$
    (C) $41 + [N] \cdot 17$
    (D) $1 + [N] \cdot 17$
    (E) none of the above

7. If $f(x) = i$, where i is an integer such that $i \le x < i + 1$, the range of $f(x)$ is

    (A) the set of all real numbers
    (B) the set of all positive integers
    (C) the set of all integers
    (D) the set of all negative integers
    (E) the set of all nonnegative real numbers

8. If $f(x) = [2x] - 4x$ with domain $0 \le x \le 2$, then $f(x)$ can also be written as

    (A) $2x$
    (B) $-x$
    (C) $-2x$
    (D) $x^2 - 4x$
    (E) none of the above

Answers and Explanations

In these solutions the following notation is used:

  i: calculator unnecessary
  a: calculator helpful or necessary
  g: graphing calculator helpful or necessary

Parametric Equations

1. g **(B)** Graph these parametric equations for values of t between -5 and 5 and for x and y between -2.5 and 2.5.

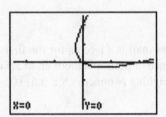

Apparently the x values are always greater than some value. Use the TRACE function to move the cursor as far left on the graph as it will go. This leads to a (correct) guess of $x \geq -\dfrac{1}{4}$. This can be verified by completing the square on the x equation:

$$x = \left(t^2 + t + \frac{1}{4}\right) - \frac{1}{4} = \left(t + \frac{1}{2}\right)^2 - \frac{1}{4}.$$

This represents a parabola that opens up with vertex at $\left(-\dfrac{1}{2}, -\dfrac{1}{4}\right)$. Therefore,

$$x \geq -\frac{1}{4}.$$

2. i **(D)** D is the only reasonable answer choice. To verify this, note that

$\dfrac{y}{2} = \cos t$. So $\dfrac{y^2}{4} = \cos^2 t$. Adding this to $x = \sin^2 t$ gives $\dfrac{y^2}{4} + x = \cos^2 t + \sin^2 t = 1$.

Since $0 \leq x \leq 1$ because $0 \leq \sin^2 t \leq 1$, this can only be a portion of the parabola given by the equation $y^2 + 4x = 4$.

3. g **(A)** You could graph all three parametric pairs to discover that only I gives a circle. (II and III give semicircles). You can also see this by a simple analysis of the equations. Removing the parameter in I by squaring and adding gives $x^2 + y^2 = 1$, which is a circle of radius 1. Substituting x for t in the y equation of II and squaring gives $x^2 + y^2 = 1$, but $y \geq 0$ so this is only a semicircle. Squaring and substituting x^2 for s in the y equation of III gives $x^2 + y^2 = 1$, but $x \geq 0$ and so this is only a semicircle.

Piecewise Functions

1. g **(B)** Enter abs($2x - 1$) into Y_1 and $4x + 5$ into Y_2. It is clear from the standard window that the two graphs intersect only at one point.

2. a **(E)** Enter abs($x - 2$) into Y_1, 1 into Y_2, and 4 into Y_3. An inspection of the graphs shows that the values of x for which the graph of Y_1 is between the other two graphs are in two intervals. E is the only answer choice having this configuration.

3. g **(A)** Subtract $|x|$ from both sides of the equation. Since $|y|$ cannot be negative, graph the piecewise function

$$Y_1 = (2 - |x|)(x \geq -2 \text{ and } x \leq 2)$$
$$Y_2 = -Y_1$$

In the first command, the word "and" is in TEST/LOGIC. The result is a square that is $2\sqrt{2}$ on a side. Therefore, the area is 8.

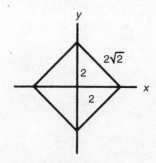

4. i **(A)** Since $f(x)$ must $= f(|x|)$, the graph must be symmetric about the y-axis. The only graph meeting this requirement is Choice A.

5. g **(B)** Since the point where a major change takes place is at $(1,1)$, the statement in the absolute value should equal zero when $x = 1$. This occurs only in Choice B. Check your answer by graphing the function in B on your graphing calculator.

6. i **(C)** If $N < 1$, and if $1 < x \leq 2$, the value should be 41; and if $1 \leq N < 2$, the value should be 58. The only answer that satisfies this is Choice C.

7. i **(C)** Since $f(x) =$ an integer by definition, the answer is Choice C.

8. g **(E)** Enter int($4x$) $- 2x$ into Y_1. The graph is shown in the figure below.

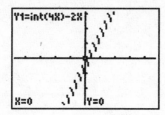

The breaks in the graph indicate that it cannot be the graph of any of the first four answer choices.

Geometry and Measurement

• Coordinate Geometry	• Three-Dimensional Geometry

2.1 Coordinate Geometry

TRANSFORMATIONS AND SYMMETRY

The three types of transformations in the coordinate plane are translations, stretches/shrinks, and reflections. Translation preserves the shape of a graph, only moving it around the coordinate plane. Translation is accomplished by addition. Changing the scale (stretching and shrinking) can change the shape of a graph. This is accomplished by multiplication. Finally, reflection preserves the shape and size of a graph but changes its orientation. Reflection is accomplished by negation.

Suppose $y = f(x)$ defines any function.

- $y = f(x) + k$ translates $y = f(x)$ k units vertically (up if $k > 0$; down if $k < 0$)
- $y = f(x - h)$ translates $y = f(x)$ h units horizontally (right if $h > 0$; left if $h < 0$)

Example 1

Suppose $y = f(x) = e^x$. Describe the graph of $y = e^x + 3$.

In this example, $k = 3$, so $e^x + 3 = f(x) + 3$. As shown in the figures below, each point on the graph of $y = e^x + 3$ is 3 units above the corresponding point on the graph of $y = e^x$.

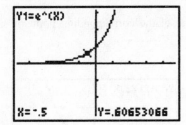
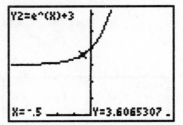

Example 2

Suppose $y = x^2$. Describe the graph of $y = (x + 2)^2$.

In this example $h = -2$, so $(x + 2)^2 = f(x + 2)$. As shown in the figures below, each point on the graph of $y = (x + 2)^2$ is 2 units to the left of the corresponding point on the graph of $y = x^2$.

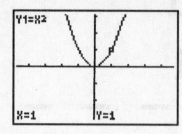 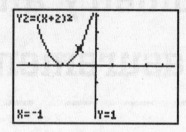

- $y = af(x)$ stretches (shrinks) $f(x)$ vertically by a factor of $|a|$ if $|a| > 1 (|a| < 1)$.

- $y = f(ax)$ shrinks (stretches) $f(x)$ horizontally by a factor of $\left|\dfrac{1}{a}\right|$ if $|a| > 1 (|a| < 1)$.

Example 3

Suppose $y = x - 1$. Describe the graph of $y = 3(x - 1)$.

In this example $|a| = 3$, so $3(|x + 1|) = 3f(x)$. As shown in the graphs below, the y-coordinate of each point on the graph of $y = 3(x - 1)$ is 3 times the y-coordinate of the corresponding point on the graph of $y = x - 1$.

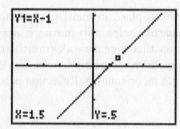 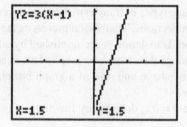

Example 4

Suppose $y = x^3$. Describe the graph of $y = \left(\dfrac{1}{2}x\right)^3$.

In this example, $|a| = \dfrac{1}{2}$. As shown in the graphs below, the x-coordinate of each point on the

graph of $y = \left(\dfrac{1}{2}x\right)^3$ is 2 times the x-coordinate of the corresponding point on the graph of

$y = x^3$.

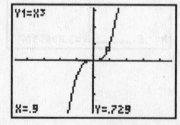 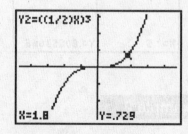

- $y = -f(x)$ reflects $y = f(x)$ about the x-axis. (The reflection is vertical.)
- $y = f(-x)$ reflects $y = f(x)$ about the y-axis. (The reflection is horizontal.)

Example 5

Suppose $y = \ln x$. Describe the graphs of $y = -\ln x$ and $y = \ln(-x)$.

As shown in the graphs below, the graph of $y = -\ln x$ is the reflection of the graph of $y = \ln x$ about the x-axis, and the graph of $y = \ln(-x)$ is the reflection about the y-axis.

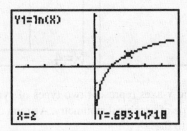

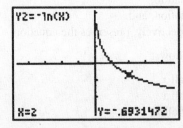 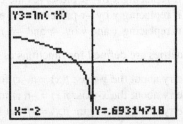

Translations, stretching/shrinking, and reflections can be combined to produce functions. Vertical transformations occur when adding, multiplying, or negating takes place after the function is applied (i.e, to y). The order in which multiple vertical transformations are executed does not matter. Horizontal transformations occur when adding, multiplying, or negating takes place before the function is applied (i.e., to x). These transformations must be taken in the following order: reflect; change the scale; then translate. Moreover, the scale factor $|a|$ must be factored out of a translation.

Example 6

Suppose $y = f(x)$. Use words to describe the transformation $y = f(-ax + b)$.

Observe that all of these transformations are horizontal. First we have to write $-ax + b$ as $-a\left(x - \dfrac{b}{a}\right)$. The x-coordinate of a point (x,y) on the graph of $y = f(x)$ goes through the following sequence of transformations: reflected about the y-axis; horizontally shrunk by a factor of $\left|\dfrac{1}{a}\right|$ or stretched by a factor of $|a|$; and translated $\dfrac{b}{a}$ to the right.

Example 7

Suppose $y = \sin x$. Describe the sequence of transformations to get the graph of $y = \sin(-2x + 1)$.

Observe that all transformations are horizontal. Write the function as $y = \sin\left(-2\left(x - \dfrac{1}{2}\right)\right)$.

Consider the point $(4, -0.7568\ldots)$ on the graph of $y = \sin x$. First reflect this point about the y-axis to $(-4, -0.7568\ldots)$. Then shrink by a factor of $\dfrac{1}{2}$ to $(-2, -0.7568\ldots)$. Then translate $\dfrac{1}{2}$ units right, to $(-1.5, -0.7568...)$. The screens below show $Y_1 = \sin x$ and $Y_2 = \sin(-2x + 1)$ and a table showing the points $(4, -0.7568...)$ for Y_1 and $(-1.5, -0.7568...)$ for Y_2.

```
Plot1 Plot2 Plot3        |  X  |  Y1   |  Y2   |
\Y1Bsin(X)               |  4  | -.7568| -.657 |
\Y2Bsin(-2X+1)■          | -1.5| -.9975| -.7568|
\Y3=                     |     |       |       |
\Y4=                     |     |       |       |
\Y5=                     |     |       |       |
\Y6=                     |     |       |       |
\Y7=                     | X=4 |       |       |
```

Reflections about the x- and y-axes represent two types of symmetry in a graph. Symmetry through the origin is a third type of graphical symmetry. A graph defined by an equation in x and y is symmetrical with respect to the

- x-axis if replacing x by $-x$ preserves the equation;
- y-axis if replacing y by $-y$ preserves the equation; and
- origin if replacing x and y by $-x$ and $-y$, respectively, preserves the equation.

These symmetries are defined for functions as follows:

- Symmetry about the y-axis: $f(x) = f(-x)$ for all x.
- Symmetry about the x-axis: $f(x) = -f(x)$ for all x.
- Symmetry about the origin: $f(x) = -f(-x)$ for all x.

As defined previously, functions that are symmetric about the y axis are even functions, and those that are symmetric about the origin are odd functions.

Example 8

Discuss the symmetry of $f(x) = \cos x$.

Since $\cos x = \cos(-x)$, cosine is symmetric about the y-axis (an even function).

However, since $\cos x \neq -\cos x$ and $\cos(-x) \neq -\cos x$, the cosine is not symmetric with respect to the x-axis or origin.

Example 9

Discuss the symmetry of $x^2 + xy + y^2 = 0$.

If you substitute $-x$ for x or $-y$ for y, but not both, the equation becomes $x^2 - xy + y^2 = 0$, which does not preserve the equation. Therefore, the graph is not symmetrical with respect to either axis. However, if you substitute both $-x$ for x and $-y$ for y, the equation is preserved, so the equation is symmetric about the origin.

EXERCISES

1. Which of the following functions transforms $y = f(x)$ by moving it 5 units to the right?

 (A) $y = f(x + 5)$
 (B) $y = f(x - 5)$
 (C) $y = f(x) + 5$
 (D) $y = f(x) - 5$
 (E) $y = 5f(x)$

2. Which of the following functions stretches $y = \cos(x)$ vertically by a factor of 3?

(A) $y = \cos(x + 3)$

(B) $y = \cos(3x)$

(C) $y = \cos\left(\dfrac{1}{3}x\right)$

(D) $y = 3 \cos x$

(E) $y = \dfrac{1}{3} \cos x$

3. The graph of $y = f(x)$ is shown.

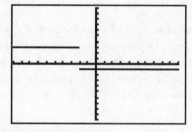

Which of the following is the graph of $y = f(-x) - 2$?

(A)

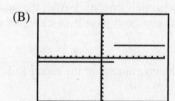

(B)

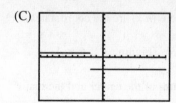

(C)

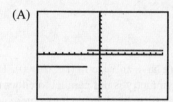

(D)

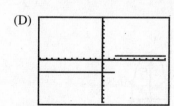

(E)

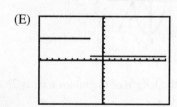

4. Which of the following is a transformation of $y = f(x)$ that translates this function down 3, shrinks it horizontally by a factor of 2, and reflects it about the x-axis.

(A) $y = -2f(x - 3)$

(B) $y = f(-2x) - 3$

(C) $y = -f\left(\frac{1}{2}x\right) - 3$

(D) $y = -f(2x) - 3$

(E) $y = 2f\left(-\frac{1}{2}x\right) - 3$

CONIC SECTIONS

The general quadratic equation in two variables has the form $Ax^2 + Bxy + Cy^2 + Dx + Ey + F = 0$, where A, B, and C are not all zero. Depending on the values of the coefficients A, B, and C, the equation represents a circle, a parabola, a hyperbola, an ellipse, or a degenerate case of one of these (e.g., a point, a line, two parallel lines, two intersecting lines, or no graph at all).

If the graph is not a degenerate case, the following indicates which conic section the equation represents:

If $B^2 - 4AC < 0$ and $A = C$, the graph is a circle.

If $B^2 - 4AC < 0$ and $A \neq C$, the graph is an ellipse.

If $B^2 - 4AC = 0$, the graph is a parabola.

If $B^2 - 4AC > 0$, the graph is a hyperbola.

The conic sections encountered on the Level 2 test will not have an xy term (i.e., $B = 0$). This term causes the graph to be rotated so that a major axis of symmetry is not parallel to either the x- or the y-axis. The general quadratic equation in two variables can be changed into a more useful form by completing the square separately on the x's and separately on the y's. After the square has been completed and, for convenience, the letters used as constants have been changed, the equation becomes:

(a) for a circle, $(x - h)^2 + (y - k)^2 = r^2$, where (h,k) are the coordinates of the center and r is the radius of the circle.

(b) for an ellipse, if $C > A$, $\dfrac{(x - h)^2}{a^2} + \dfrac{(y - k)^2}{b^2} = 1$, where (h,k) are the coordinates

of the center and the major axis is parallel to the x-axis (Figure a). If $C < A$,

$\dfrac{(x - h)^2}{b^2} + \dfrac{(y - k)^2}{a^2} = 1$, where (h,k) are the coordinates of the center and the major

axis is parallel to the y-axis (Figure b).

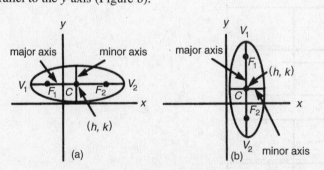

In both cases the length of the *major axis* is $2a$ and the length of the *minor axis* is $2b$.

(c) for a parabola, if $C = 0$, $(x - h)^2 = 4p(y - k)$ (Figure a). If $A = 0$, $(y - k)^2 = 4p(x - h)$ (Figure b). In both cases (h,k) are the coordinates of the vertex and p is the directed distance from the vertex to the focus.

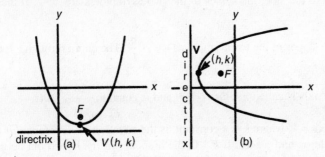

(d) for a hyperbola, $\dfrac{(x - h)^2}{a^2} - \dfrac{(y - k)^2}{b^2} = 1$, where (h,k) are the coordinates of the center,

and the graph opens to the side (Figure a). If the graph opens up and down, the equation

is $\dfrac{(y - k)^2}{a^2} - \dfrac{(x - h)^2}{b^2} = 1$, where (h,k) are the coordinates of the center (Figure b).

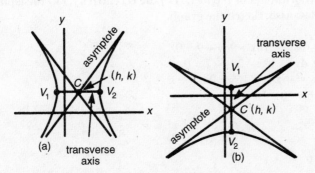

In both cases the length of the *transverse axis* is $2a$, and the length of the *conjugate axis* is $2b$.

An *ellipse* is the set of points in a plane whose distances from two fixed points, called *foci,* **sum to** a constant, equal to $2a$ in the formula on the previous page. The distance between the center and a focus is called c and is equal to $\sqrt{a^2 - b^2}$. The *eccentricity* of an ellipse is always less than 1 and is equal to $\dfrac{c}{a}$. The chord through a focus perpendicular to the major axis is called a *latus rectum* and is $\dfrac{2b^2}{a}$ units long.

A *parabola* is the set of points in a plane whose distances from a fixed point, called the *focus,* and a fixed line, called the *directrix* are equal. The *eccentricity* is always equal to 1. The chord through the focus perpendicular to the axis of symmetry is called the *latus rectum* and is $4p$ units long.

A *hyperbola* is the set of points in a plane whose distances from two fixed points, called *foci,* **differ** by a constant, equal to $2a$ in the formula previously stated. The distance between the center and a focus is called c and is equal to $\sqrt{a^2 + b^2}$. The *eccentricity* of a hyperbola is always greater than 1 and is equal to $\dfrac{c}{a}$. Every hyperbola has associated with it two lines, called *asymptotes,* that intersect at the center. In general, an asymptote is a line that a curve approaches, but never quite touches, as one or both variables become increasingly larger or smaller. If the hyperbola opens to the side, the slopes of the two asymptotes are $\pm\dfrac{b}{a}$. If the hyperbola opens up and down, the slopes of the twoasymptotes are $\pm\dfrac{a}{b}$. The chord through a focus perpendicular to the transverse axis is called a *latus rectum* and is equal to $\dfrac{2b^2}{a}$ units.

The equation $xy = k$, where k is a constant, is the equation of a *rectangular hyperbola* whose asymptotes are the x- and y-axes. If $k > 0$, the branches of the hyperbola lie in quadrants I and III. If $k < 0$, the branches lie in quadrants II and IV.

Example 1

Each of the following is an equation of a conic section. State which one and find, if they exist: (I) the coordinates of the center, (II) the coordinates of the vertices, (III) the coordinates of the foci, (IV) the eccentricity, (V) the equations of the asymptotes. Also, sketch the graph.

(A) $9x^2 - 16y^2 - 18x + 96y + 9 = 0$
(B) $4x^2 + 4y^2 - 12x - 20y - 2 = 0$
(C) $4x^2 + y^2 + 24x - 16y = 0$
(D) $y^2 + 6x - 8y + 4 = 0$

SOLUTION

In all cases $B = 0$.

(A) $B^2 - 4AC = 0 - 4(9)(-16) > 0$, which means the graph will be a hyperbola. To complete the square, group the x-terms and the y-terms. Factor out of each group the coefficient of the quadratic term:

$$9(x^2 - 2x \quad) - 16(y^2 - 6y \quad) + 9 = 0$$

A perfect trinomial square is obtained within both parentheses by taking one-half of the coefficient of the linear term, squaring it, and adding it to the existing polynomial. A similar amount must be added to the right side of the equation.

$$9(x^2 - 2x + 1) - 16(y^2 - 6y + 9) + 9 = 9 \cdot 1 - 16 \cdot 9$$
$$9(x - 1)^2 - 16(y - 3)^2 = -144$$

Divide each term by -144 to get the equation in the graphing form: $\dfrac{(y-3)^2}{9} - \dfrac{(x-1)^2}{16} = 1$.

  I. Center $(1,3)$. $a = 3$, $b = 4$, and so $c = \sqrt{9 + 16} = 5$.
 II. The vertices are 3 units above and 3 units below the center. Vertices are $(1,6)$ and $(1,0)$.
III. The foci are 5 units above and 5 units below the center. Foci are $(1,8)$ and $(1,-2)$.
 IV. Eccentricity $= \dfrac{c}{a} = \dfrac{5}{3}$.

V. The slopes of the asymptotes are $\pm\dfrac{a}{b}=\pm\dfrac{3}{4}$. The asymptotes pass through the center

(1,3). Therefore, the equations of the asymptotes are $y-3=\pm\dfrac{3}{4}(x-1)$.

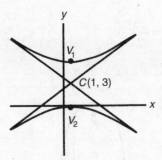

(B) $B^2-4AC=0-4\cdot4\cdot4<0$ and $A=C=4$, which means the graph will be a circle. Divide through by 4, group the x-terms together, group the y-terms together, and complete the square.

$$\left(x^2-3x+\frac{9}{4}\right)+\left(y^2-5y+\frac{25}{4}\right)-\frac{2}{4}=\frac{9}{4}+\frac{25}{4}$$

$$\left(x-\frac{3}{2}\right)^2+\left(y-\frac{5}{2}\right)^2=9$$

I. Center $\left(\dfrac{3}{2},\dfrac{5}{2}\right)$. Radius = 3. None of the other items is defined.

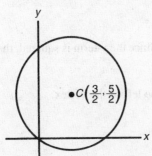

(C) $B^2-4AC=0-4\cdot4\cdot1<0$ and $A\neq C$, which means the graph will be an ellipse. Group the x-terms, group the y-terms, and factor out of each group the coefficient of the quadratic term:

$$4(x^2+6x\quad)+1(y^2-16y\quad)=0$$

Complete the square and add similar amounts to both sides of the equation:

$$4(x^2+6x+9)+1(y^2-16y+64)=4\cdot9+1\cdot64$$
$$4(x+3)^2+(y-8)^2=100$$

Divide each term by 100 to get the equation in the proper form:

$$\frac{(x+3)^2}{25}+\frac{(y-8)^2}{100}=1$$

I. Center $(-3,8)$. $a = 10$, $b = 5$, and so $c = \sqrt{100 - 25} = \sqrt{75} = 5\sqrt{3}$.

II. Since the major axis is vertical, the vertices are 10 units above and 10 units below the center. Vertices are $(-3,18)$ and $(-3,-2)$.

III. The foci are $5\sqrt{3}$ units above and $5\sqrt{3}$ units below the center. Foci are $(-3,8 + 5\sqrt{3})$ and $(-3,8 - 5\sqrt{3})$.

IV. Eccentricity $= \dfrac{c}{a} = \dfrac{5\sqrt{3}}{10} = \dfrac{\sqrt{3}}{2}$.

V. There are no asymptotes.

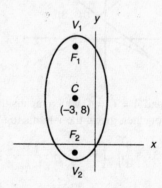

(D) $B^2 - 4AC = 0 - 4 \cdot 0 \cdot 1 = 0$, which means the graph will be a parabola. Group the y-terms and complete the square.

$$(y^2 - 8y + 16) = -6x - 4 + 16$$
$$(y - 4)^2 = -6(x - 2)$$

I. No center.

II. Vertex is $(2,4)$.

III. $4p = -6$, and so $p = -\dfrac{3}{2}$. Since the y-term is squared, the parabola opens to the side. Thus, the focus is $\dfrac{3}{2}$ units to the left of the vertex.

IV. Eccentricity $= 1$.

V. No asymptotes.

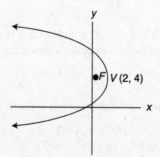

![Example 2]

Find the equation of the hyperbola with center at (3,–4), eccentricity 4, and conjugate axis of length 6.

Two cases are possible: $\dfrac{(x-3)^2}{a^2} - \dfrac{(y+4)^2}{b^2} = 1$ and $\dfrac{(y+4)^2}{a^2} - \dfrac{(x-3)^2}{b^2} = 1$.

Eccentricity $= \dfrac{c}{a} = 4$. Thus, $c = 4a$. $\left(\dfrac{1}{2}, 4\right)$

Conjugate axis $= 2b = 6$. Thus, $b = 3$.
In the case of a hyperbola,

$$c^2 = a^2 + b^2$$
$$(4a)^2 = a^2 + 9$$
$$16a^2 = a^2 + 9$$
$$15a^2 = 9$$
$$a^2 = \frac{9}{15}$$

Therefore, the equation becomes:

$$\frac{15(x-3)^2}{9} - \frac{(y+4)^2}{9} = 1 \text{ or } \frac{15(y+4)^2}{9} - \frac{(x-3)^2}{9} = 1.$$

A *degenerate case* occurs when one of the variables drops out, or the terms with the variables equal zero, or a negative number after the square has been completed.

Example 3

What is the graph of each of the following?

(A) $x^2 + y^2 - 4x + 2y + 5 = 0$
(B) $xy = 0$
(C) $2x^2 - 3y^2 + 8x + 6y + 5 = 0$
(D) $3x^2 + 4y^2 - 6x - 16y + 19 = 0$
(E) $x^2 + y^2 + 5 = 0$

SOLUTIONS

(A) $B^2 - 4AC = 0 - 4 \cdot 1 \cdot 1 < 0$ and $A = C$, which indicates the graph should be a circle. Completing the square on the x's and y's separately gives:

$$(x^2 - 4x + 4) + (y^2 + 2y + 1) + 5 = 0 + 4 + 1$$
$$(x - 2)^2 + (y + 1)^2 = 0$$

This is a circle with center $(2,-1)$ and radius zero; thus, a degenerate case. The graph is the single point $(2,-1)$.

(B) $xy = 0$ if and only if $x = 0$ or $y = 0$. Thus, the graph is two intersecting lines: the x-axis ($y = 0$) and the y-axis ($x = 0$).

(C) $B^2 - 4AC = 0 - 4 \cdot 2 \cdot (-3) > 0$, which indicates the graph should be a hyperbola. Factoring and completing the square on the x's and y's separately gives:

$$2(x^2 + 4x + 4) - 3(y^2 - 2y + 1) + 5 = 0 + 2 \cdot 4 - 3 \cdot 1$$
$$2(x + 2)^2 - 3(y - 1)^2 = 0$$

Since the right side of the equation is zero and the left side is the difference between two perfect squares, the left side of the equation can be factored and simplified.

$$\left[\sqrt{2}(x+2)+\sqrt{3}(y-1)\right]\cdot\left[\sqrt{2}(x+2)-\sqrt{3}(y-1)\right]=0$$

$$\left[\sqrt{2}x+2\sqrt{2}+\sqrt{3}y-\sqrt{3}\right]=0$$

or $$\left[\sqrt{2}x+2\sqrt{2}-\sqrt{3}y+\sqrt{3}\right]=0$$

$$y=-\frac{\sqrt{2}}{\sqrt{3}}x+\frac{\sqrt{3}-2\sqrt{2}}{\sqrt{3}}\approx-0.82x-0.63$$

or $$y=\frac{\sqrt{2}}{\sqrt{3}}x+\frac{\sqrt{3}+2\sqrt{2}}{\sqrt{3}}\approx0.82x+2.63$$

Thus, the graph is two lines intersecting at (–2,1), one with slope $-\dfrac{\sqrt{2}}{\sqrt{3}}\approx-0.82$ and y-intercept $\dfrac{\sqrt{3}-2\sqrt{2}}{\sqrt{3}}\approx-0.63$ and the other with slope $\dfrac{\sqrt{2}}{\sqrt{3}}\approx0.82$ and y-intercept $\dfrac{\sqrt{3}+2\sqrt{2}}{\sqrt{3}}\approx2.63$.

(D) $B^2-4AC=0-4\cdot3\cdot4<0$ and $A\neq C$, which indicates the graph should be an ellipse. Factoring and completing the square on the x's and y's separately gives:

$$3(x^2-2x+1)+4(y^2-4y+4)+19=0+3\cdot1+4\cdot4$$
$$3(x-1)^2+4(y-2)^2=0$$

Since $3(x-1)^2\geq0$ and $4(y-2)^2\geq0$, the only point that satisfies the equation is (1,2), which makes each term on the left side of the equation equal to zero. The graph is the one point (1,2).

(E) $x^2+y^2=-5$. The graph does not exist since $x^2\geq0$ and $y^2\geq0$ and there is no way that their sum can equal –5.

EXERCISES

1. Which of the following are the coordinates of a focus of $5x^2+4y^2-20x+8y+4=0$?

    (A) (1,–1)
    (B) (2,–1)
    (C) (3,–1)
    (D) (2,–2)
    (E) (–2,1)

2. If the graphs of $x^2+y^2=4$ and $xy=1$ are drawn on the same set of axes, how many points will they have in common?

    (A) 0
    (B) 1
    (C) 2
    (D) 3
    (E) 4

3. The graph of $(x - 2)^2 = 4y$ has a

   (A) vertex at (4,2)
   (B) focus at (2,0)
   (C) directrix $y = -1$
   (D) latus rectum 2 units in length
   (E) none of these

4. Which of the following is an asymptote of $3x^2 - 4y^2 - 12 = 0$?

   (A) $y = \dfrac{4}{3}x$

   (B) $y = -\dfrac{2}{\sqrt{3}}$

   (C) $y = -\dfrac{3}{4}x$

   (D) $y = \dfrac{\sqrt{3}}{2}x$

   (E) $y = \dfrac{2\sqrt{3}}{3}x$

5. The graph of $x^2 = (2y + 3)^2$ is

   (A) a circle
   (B) an ellipse
   (C) a hyperbola
   (D) a point
   (E) two intersecting lines

6. The area bounded by the curve $y = \sqrt{4 - x^2}$ and the x-axis is

   (A) 4π
   (B) 8π
   (C) 16π
   (D) 2π
   (E) π

7. An equilateral triangle is inscribed in the circle whose equation is $x^2 + 2x + y^2 - 4y = 0$.
   The length of the side of the triangle is

   (A) 5
   (B) 1.9
   (C) 2.2
   (D) 3.9
   (E) 4.5

POLAR COORDINATES

Although the most common way to represent a point in a plane is in terms of its distances from two perpendicular axes, there are several other ways. One such way is in terms of the distance of the point from the origin and the angle between the positive *x*-axis and the ray emanating from the origin going through the point.

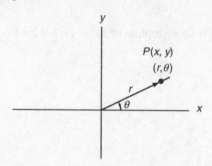

In the figure, the regular rectangular coordinates of *P* are (*x,y*) and the *polar coordinates* are (*r,θ*). Note that $r > 0$ because it represents a distance.

Since $\sin\theta = \dfrac{y}{r}$ and $\cos\theta = \dfrac{x}{r}$, there is an easy relationship between rectangular and polar coordinates:

$x = r\cos\theta$
$y = r\sin\theta$
$x^2 + y^2 = r^2$, using the Pythagorean theorem.

Unlike the case involving rectangular coordinates, each point in the plane can be named by an infinite number of polar coordinates.

Example 1

(2,30°), (2,390°), (2,–330°), (–2,210°), (–2,–150°) all name the same point.

In general, a point in the plane represented by (*r,θ*) can also be represented by $(r,\theta + 2\pi n)$ or $(-r,\theta + (2n - 1)\pi)$, where *n* is an integer.

Example 2

Express point *P*, whose rectangular coordinates are $\left(3, 3\sqrt{3}\right)$, in terms of polar coordinates.

$$r^2 = x^2 + y^2 = 9 + 27 = 36$$
$$r = 6$$
$$r\cos\theta = x$$
$$\cos\theta = \frac{3}{6} = \frac{1}{2}$$

Therefore, $\theta = 60°$, and the coordinates of *P* are (6,60°).

Example 3

Describe the graphs of (A) $r = 2$ and (B) $r = \dfrac{1}{\sin\theta}$.

(A) $r^2 = x^2 + y^2$

$\qquad r = 2$

Therefore, $x^2 + y^2 = 4$, which is the equation of a circle whose center is at the origin and whose radius is 2.

(B) $r \sin\theta = y$

$\qquad r = \dfrac{1}{\sin\theta}$

Therefore, $y = 1$.

Thus, $r = \dfrac{1}{\sin\theta}$ is the equation of a horizontal line one unit above the *x*-axis.

Both $r = 2$ and $r = \dfrac{1}{\sin\theta}$ are examples of polar functions. Such functions can be graphed on a graphing calculator by using POLAR mode. Enter 2 into r_1 and graph using ZOOM 6, the standard window.

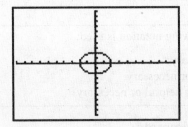

Although the graph is accurate for the scale given, the shape looks like that of an ellipse, not a circle. This is because the standard screen has a scale on the *x*-axis that is larger than that on the *y*-axis, resulting in a distorted graph. The graph below is obtained by using the ZOOM ZSquare command.

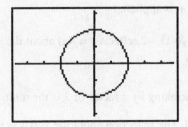

EXERCISES

1. A point has polar coordinate $(2, 60°)$. The same point can be represented by

   (A) $(-2, 240°)$
   (B) $(2, 240°)$
   (C) $(-2, 60°)$
   (D) $(2, -60°)$
   (E) $(2, -240°)$

2. The polar coordinates of a point P are $(2, 200°)$. The rectangular coordinates of P are

    (A) $(-1.88, -0.68)$
    (B) $(-0.68, -1.88)$
    (C) $(-0.34, -0.94)$
    (D) $(-0.94, -0.34)$
    (E) $(-0.47, -0.17)$

3. Describe the graph of $r = \dfrac{3}{\cos\theta}$.

    (A) a parabola
    (B) an ellipse
    (C) a circle
    (D) a vertical line
    (E) the x axis

Answers and Explanations

In these solutions the following notation is used:

    i: calculator unnecessary
    a: calculator helpful or necessary
    g: graphing calculator helpful or necessary

Transformations and Symmetry

1. i   **(B)** Horizontal translation (right) is accomplished by subtracting the amount of the translation (5) from x before the function is applied.

2. i   **(D)** Vertical stretching is accomplished by multiplying the function by the stretching factor after the function is applied.

3. i   **(D)** The graph of $y = f(-x) - 2$ reflects $y = f(x)$ about the y axis and translates it down 2.

4. i   **(C)** The horizontal shrinking by a factor of 2 is the multiplication of x by $\dfrac{1}{2}$ before the function is applied. The reflection about the x-axis is the negation of the function after it is applied. The translation down 3 is the addition of -3 after the function is applied.

Conic Sections

1. i   **(B)** Complete the squares: $5(x^2 - 4x + 4) + 4(y^2 + 2y + 1) = -4 + 20 + 4$.

    $5(x-2)^2 + 4(y+1)^2 = 20$. $\dfrac{(x-2)^2}{4} + \dfrac{(y+1)^2}{5} = 1$. $a^2 = 5$, $b^2 = 4$, and so $c^2 = 1$.

    Therefore, the foci are 1 unit above and below the center, which is at $(2, -1)$.

2. i **(E)** See the figure below.

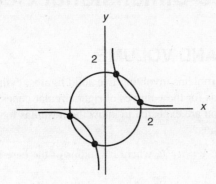

3. i **(C)** Vertex is at $(2,0)$, focus is at $(2,1)$, directrix is $y = -1$.

4. i **(D)** $\dfrac{x^2}{4} - \dfrac{y^2}{3} = 1$. Asymptotes $y = \pm\dfrac{\sqrt{3}}{2}x$.

5. i **(E)** $x^2 - (2y + 3)^2 = 0$ factors into $(x - 2y - 3)(x + 2y + 3) = 0$, which breaks into $x - 2y - 3 = 0$ or $x + 2y + 3 = 0$. These are the equations of two intersecting lines.

6. i **(D)** This is the equation of a semicircle with radius 2. $A = \dfrac{1}{2}\pi r^2 = 2\pi$.

7. a **(D)** This is an equation of a circle. Complete the square to determine that the radius is $\sqrt{5}$. Since $\triangle ABC$ is an equilateral triangle, arc $AB = 120°$. Central angle $\angle AOB = 120°$, OM is an altitude, and $\angle AOM = 60°$. Thus, $\triangle AOM$ is a 30-60-90 triangle. Therefore, $AM = \dfrac{\sqrt{5}}{2}\sqrt{3}$, so $AB = \sqrt{15} \approx 3.9$.

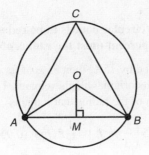

Polar Coordinates

1. i **(A)** The angle must either be coterminal with $(60 \pm 360n)$ with $r = 2$ or $(60 \pm 180)°$ with $r = -2$. A is the only answer choice that meets these criteria.

2. a **(A)** With your calculator in degree mode, evaluate $x = r\cos\theta = 2\cos 200 \approx -1.88$ and $y = r\sin 200 = 2\sin 200 \approx -0.68$.

3. g **(D)** With your graphing calculator in POLAR mode, enter $\dfrac{3}{\cos\theta}$ as r_1, and observe that the graph is a vertical line.

2.2 Three-Dimensional Geometry

SURFACE AREA AND VOLUME

The Level 2 test may have problems involving five solid figures: cylinders, cones, pyramids, spheres, and prisms. Formulas for the volumes of right circular cones, spheres, and pyramids are given in the test book. You are expected to know the formulas for the volumes of right circular cylinders and right prisms:

- Right circular cylinders: $V = \pi r^2 h$, where r = radius of the base and h = height of the cylinder
- Right prisms: $V = lwh$, where l = length, w = width, and h = height of the prism

Formulas for the lateral area of a right circular cone and the surface area of a sphere are also given. Since the sides of a pyramid are triangles and the sides of a prism are rectangles, you are expected to know how to find these areas as well.

Problems on these solid figures also appear on the Level 1 test. The problems on the Level 2 test are generally more difficult.

Example 1

The length, width, and height of a right prism are 9, 4, and 2, respectively. What is the length of the longest segment whose endpoints are vertices of the prism?

The longest such segment is a diagonal of the prism. The length of this diagonal is

$$l = \sqrt{9^2 + 4^2 + 2^2} = \sqrt{101} \,.$$

Example 2

If the volume of a right circular cone is to be reduced by 15% by reducing its height by 5%, by what percent must the radius of the base be reduced?

If the volume of the cone is reduced by 15%, then 85% of its original volume remains. If its height is reduced by 5%, 95% of its original height remains. Use the formula for the volume of a cone, and let p be the proportion of the radius that remains,

$$0.85 \left(\frac{1}{3} \pi r^2 h \right) = \frac{1}{3} \pi (pr)^2 (0.95h)$$

This equation simplifies to $0.85 = 0.95p^2$, so $p \approx 0.946$. Therefore, the radius must be reduced by 5.4%.

A solid figure can be obtained by rotating a plane figure about some line in the plane that does not intersect the figure.

Example 3

If the segment of the line $y = -2x + 2$ that lies in quadrant I is rotated about the y-axis, a cone is formed. What is the volume of the cone?

As shown in the figure below on the left, the part of the segment that lies in the first quadrant and the axes form a triangle with vertices at $(0,0)$, $(1,0)$, and $(0,2)$. Rotating this triangle about the y-axis generates the cone shown in the figure below on the right.

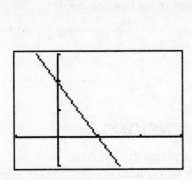 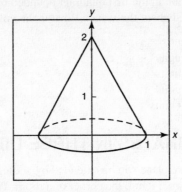

The radius of the base is 1, and the height is 2. Therefore, the volume is

$$V = \frac{1}{3}\pi(1)^2(2) = \frac{2}{3}\pi .$$

EXERCISES

1. The figure below shows a right circular cylinder inscribed in a cube with edge of length x. What is the ratio of the volume of the cylinder to the volume of the cube?

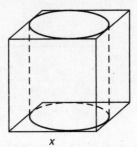

x

(A) $\dfrac{2}{3}$

(B) $\dfrac{3}{4}$

(C) $\dfrac{\pi}{4}$

(D) $\dfrac{\pi}{2}$

(E) $\dfrac{4}{5}$

2. The volume of a right circular cylinder is the same numerical value as its total surface area. Find the smallest integral value for the radius of the cylinder.

(A) 1
(B) 2
(C) 3
(D) 4
(E) This value cannot be determined.

3. The region in the first quadrant bounded by the line $3x + 2y = 7$ and the coordinate axes is rotated about the x-axis. What is the volume of the resulting solid?

(A) 8 units^3
(B) 20 units^3
(C) 30 units^3
(D) 90 units^3
(E) 120 units^3

COORDINATES IN THREE DIMENSIONS

The coordinate plane can be extended by adding a third axis, the z-axis, which is perpendicular to the other two. Picture the corner of a room. The corner itself is the origin. The edges between the walls and the floor are the x- and y-axes. The edge between the two walls is the z-axis. The first octant of this three-dimensional coordinate system and the point (1,2,3) are illustrated below. The cone shown in the figure above extends into three-dimensional space and could be located by using a z-axis that is perpendicular to the paper.

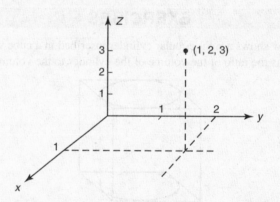

A point that has zero as any coordinate must lie on the plane formed by the other two axes. If two coordinates of a point are zero, then the point lies on the nonzero axis. The three-dimensional Pythagorean theorem yields a formula for the distance between two points (x_1,y_1,z_1) and (x_2,y_2,z_2) in space:

$$d = \sqrt{(x_2 - x_1)^2 + (y_2 - y_2)^2 + (z_2 - z_1)^2}$$

A three-dimensional coordinate system can be used to graph the variable z as a function of the two variables x and y, but such graphs are beyond the scope of the Level 2 Test.

EXERCISES

1. The distance between two points in space, $P(x,-1,-1)$ and $Q(3,-3,1)$, is 3. Find the possible values of x.

   (A) 1 or 2
   (B) 2 or 3
   (C) −2 or −3
   (D) 2 or 4
   (E) −2 or −4

   $$\sqrt{(x-3)^2+(-1+3)^2+(-1+1)^2}=\sqrt{(x-3)^2+8}=3$$
   $$(x-3)^2=1$$
   $$x-3=\pm1$$
   $$x=4,2$$

2. The point $(-4,0,7)$ lies on the

   (A) y axis
   (B) xy plane
   (C) yz plane
   (D) xz plane
   (E) z axis

Answers and Explanations

In these solutions the following notation is used:

> i: calculator unnecessary
> a: calculator helpful or necessary
> g: graphing calculator helpful or necessary

Surface Area and Volume

1. i **(C)** The volume of the cube is x^3. The radius of the cylinder is $\dfrac{x}{2}$, and its height is x.

   Substitute these into the formula for the volume of a cylinder:

   $$V = \pi r^2 h$$
   $$= \pi\left(\frac{x}{2}\right)^2 x$$
   $$= \frac{\pi}{4}x^3$$

2. i **(C)** $V = \pi r^2 h$, and total area $= 2\pi r^2 + 2\pi rh$. Setting these two equal yields $rh = 2r + 2h$.

   Therefore, $h = \dfrac{2r}{r-2}$. Since h must be positive, the smallest integer value of r is 3.

3. a **(C)** The line $3x + 2y = 7$ has x-intercept $\dfrac{7}{3}$ and y-intercept $\dfrac{7}{2}$. The part of this line that lies in the first quadrant forms a triangle with the coordinate axes. Rotating this triangle about the x-axis produces a cone with radius $\dfrac{7}{2}$ and height $\dfrac{7}{3}$. The volume of this cone is $\dfrac{1}{3}\pi\left(\dfrac{7}{2}\right)^2\left(\dfrac{7}{3}\right) \approx 30$.

Coordinates in Three Dimensions

1. i **(D)** The square of the distance between P and Q is 9, so

$$(x-3)^2 + (-1-(-3))^2 + (-1-1)^2 = 9, \text{ or } (x-3)^2 = 1.$$

Therefore, $x - 3 = \pm 1$, so $x = 2$ or 4.

2. i **(D)** Since the y coordinate is zero, the point must lie in the xz plane.

Numbers and Operations

CHAPTER 3

- Counting
- Complex Numbers
- Matrices
- Sequences and Series
- Vectors

3.1 Counting

VENN DIAGRAMS

Counting problems usually begin with the phrase "How many . . ." or the phrase "In how many ways . . ." Illustrating counting techniques by example is best.

Example 1

A certain sports club has 50 members. Of these, 35 golf, 30 hunt, and 18 do both. How many club members do neither?

Add 35 and 30, then subtract the 18 that were counted twice. This makes 47 who golf, hunt, or do both. Therefore, only 3 (50 – 47) do neither.

Example 2

Among the seniors at a small high school, 80 take math, 41 take Spanish, and 54 take physics. Ten seniors take math and Spanish; 19 take math and physics; and 12 take physics and Spanish. Seven seniors take all three. How many seniors take math but not Spanish or physics?

A Venn diagram will help you sort out this complicated-sounding problem.

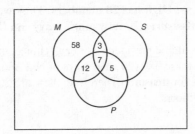

Start with the 7 who take all three courses. Since 10 take math and Spanish, this leaves 3 who take math and Spanish but not physics. Use similar reasoning to see that 5 take physics and Spanish but not math, and 12 take math and physics but not Spanish.

Finally, using the totals for how many students take each course, we can conclude that 58 (80 − 3 − 7 − 12) take math but not physics or Spanish.

EXERCISE

1. There are 50 people in a room. Twenty-eight are male, and 32 are under the age of 30. Twelve are males under the age of 30. How many women over the age of 30 are in the group?

   (A) 2
   (B) 3
   (C) 4
   (D) 5
   (E) 6

MULTIPLICATION RULE

Many other counting problems use the multiplication principle.

Example 1

Suppose you have 5 shirts, 4 pairs of pants, and 9 ties. How many outfits can be made consisting of a shirt, a pair of pants, and a tie?

For each of the 5 shirts, you can wear 4 pairs of pants, so there are $5 \cdot 4 = 20$ shirt-pants combinations. For each of these 20 shirt-pants combinations, there are 9 ties, so there are $20 \cdot 9 = 180$ shirt-pants-tie combinations.

Example 2

Six very good friends decide they will have lunch together every day. In how many different ways can they line up in the lunch line?

Any one of the 6 could be first in line. For each person who is first, there are 5 who could be second. This means there are 30 ($6 \cdot 5$) ways of choosing the first two people. For each of these 30 ways, there are 4 ways of choosing the third person. This makes 120 ($30 \cdot 4$) ways of choosing the first 3 people. Continuing in this fashion, there are $6 \cdot 5 \cdot 4 \cdot 3 \cdot 2 \cdot 1 = 720$ ways these 6 friends can stand in the cafeteria line. (This means that if they all have perfect attendance for 4 years of high school, they could stand in line in a different order every day, because $720 = 4 \cdot 180$.)

Example 3

The math team at East High has 20 members. They want to choose a president, vice president, and treasurer. In how many ways can this be done?

Any one of the 20 members could be president. For each choice, there are 19 who could be vice president. For each of these 380 ($20 \cdot 19$) ways of choosing a president and a vice president, there are 18 choices for treasurer. Therefore, there are $20 \cdot 19 \cdot 18 = 6840$ ways of choosing these three club officers.

Example 4

The student council at West High has 20 members. They want to select a committee of 3 to work with the school administration on policy matters affecting students directly. How many committees of 3 students are possible?

This problem is similar to example 3, so we will start with the fact that if they were electing 3 officers, the student council would be able to do this in 6840 ways. However, it does not matter whether member A is president, B is vice president, and C is treasurer or some other arrangement, as long as all 3 are on the committee. Therefore, we can divide 6840 by the number of ways the 3 students selected could be president, vice president, and treasurer. This latter number is $3 \cdot 2 \cdot 1 = 6$, so there are 1140 ($6840 \div 6$) committees of 3.

EXERCISE

1. M & M plain candies come in six colors: brown, green, orange, red, tan, and yellow. Assume there are at least 3 of each color. If you pick three candies from a bag, how many color possibilities are there?

    (A) 18
    (B) 20
    (C) 120
    (D) 216
    (E) 729

FACTORIAL, PERMUTATIONS, COMBINATIONS

Counting problems like the ones in the last three examples occur frequently enough that they have special designations.

Ordering *n* Objects (Factorial)

The second example of the multiplication rule asked for the number of ways 6 friends could stand in line. By using the multiplication principle, we found that there were $6 \cdot 5 \cdot 4 \cdot 3 \cdot 2 \cdot 1$ ways. A special notation for this product is 6! (6 factorial). In general, the number of ways objects can be ordered is $n!$

Ordering *r* of *n* Objects (Permutations)

The third example of the multiplication rule asked for the number of ways you could choose a first (president), second (vice president), and third person (secretary) out of 20 people ($r = 3$, $n = 20$). The answer is $20 \cdot 19 \cdot 18$, or $\dfrac{20!}{17!} = \dfrac{20!}{(20-3)!}$. In general, there are $\dfrac{n!}{(n-r)!}$ permutations of *r* objects of *n*. This appears as ${}_nP_r$ in the calculator menu.

Choosing *r* of *n* Objects (Combinations)

In the fourth example of the multiplication rule, we were interested in choosing a committee of 3 where there was no distinction among committee members. Our approach was first to compute the number of ways of choosing officers and then dividing out the number of ways the three officers could hold the different offices. This led to the computation $\dfrac{20!}{17!3!} = \dfrac{20!}{(20-3)!3!}$. In general, the number of ways of choosing r of n objects is $\dfrac{n!}{(n-r)!r!}$. This quantity appears on

the calculator menu as $_nC_r$. However, there is a special notation for combinations: $_nC_r = \begin{pmatrix} n \\ r \end{pmatrix} =$ the number of ways r objects can be chosen from n.

Calculator commands for all three of these functions are in the MATH/PRB menu.

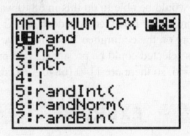

These 3 commands can also be found on scientific calculators.

EXERCISES

1. How many 3-person committees can be selected from a fraternity with 25 members?

   (A) 15,625
   (B) 13,800
   (C) 2,300
   (D) 75
   (E) 8

2. A basketball team has 5 centers, 9 guards, and 13 forwards. Of these, 1 center, 2 guards, and 2 forwards start a game. How many possible starting teams can a coach put on the floor?

   (A) 56,160
   (B) 14,040
   (C) 585
   (D) 197
   (E) 27

3. Five boys and 6 girls would like to serve on the homecoming court, which will consist of 2 boys and 2 girls. How many different homecoming courts are possible?

   (A) 30
   (B) 61
   (C) 150
   (D) 900
   (E) 2048

4. In a plane there are 8 points, no three of which are collinear. How many lines do the points determine?

   (A) 7
   (B) 16
   (C) 28
   (D) 36
   (E) 64

5. If $\begin{pmatrix} 6 \\ x \end{pmatrix} = \begin{pmatrix} 4 \\ x \end{pmatrix}$, then $x =$

 (A) 0
 (B) 1
 (C) 4
 (D) 5
 (E) 10

Answers and Explanations

In these solutions the following notation is used:

   i: calculator unnecessary
   a: calculator helpful or necessary
   g: graphing calculator helpful or necessary

Venn Diagrams

1. i   **(A)** Venn diagram will help you solve this problem.

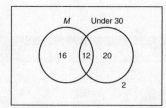

The two circles represent males and people who are at most 30 years of age, respectively. The part of the rectangle outside both circles represents people who are in neither category, i.e., females over the age of 30. First fill in the 12 males who are less than or equal to 30 years of age in the intersection of the circles. Since there are 28 males altogether, 16 are male and over 30. Since there are 32 people age 30 or less, there are 20 women that age. Add these together to get 48 people. Since there are 50 in the group, 2 must be women over 30.

Multiplication Rule

1. a   **(D)** There are 6 choices of color for each of the three candies selected. Therefore, there are $6 \times 6 \times 6 = 216$ color possibilities altogether.

Factorial, Permutations, Combinations

1. a   **(C)** This is the number of ways 3 objects can be chosen from 25, or $\begin{pmatrix} 25 \\ 3 \end{pmatrix} =$ $25_nC_r3 = 2{,}300$.

2. a   **(B)** There are $\begin{pmatrix} 5 \\ 1 \end{pmatrix} = 5$ ways of choosing the one center, $\begin{pmatrix} 9 \\ 2 \end{pmatrix} = 36$ ways of choosing the two guards, and $\begin{pmatrix} 13 \\ 2 \end{pmatrix} = 78$ ways of choosing the two forwards. Therefore there are $5 \times 36 \times 78 = 14{,}040$ possible starting teams.

3. a **(C)** There are $\binom{5}{2} = 10$ ways of choosing 2 boys out of 5 and $\binom{6}{2} = 15$ ways of choosing 2 girls out of 6. Therefore there are $10 \times 15 = 150$ ways of choosing the homecoming court.

4. a **(C)** Since no three points are collinear, every pair of points determines a distinct line. There are $\binom{8}{2} = 28$ such lines.

5. i **(A)** $\binom{n}{0} = 1$ for any n.

3.2 Complex Numbers

IMAGINARY NUMBERS

The square of a real number is never negative. This means that the square root of a negative number cannot be a real number. The symbol $i = \sqrt{-1}$ is called the imaginary unit, $i^2 = -1$. Powers of i follow a pattern:

Power of i	Intermediate Steps	Value
i^1	i	i
i^2	$i \cdot i = -1$	-1
i^3	$i^2 \cdot i = (-1) \cdot i = -i$	$-i$
i^4	$i^3 \cdot i = (-i) \cdot i = -i^2 = -(-1) = 1$	1
i^5	$i^4 \cdot i = 1 \cdot i = i$	i

In other words, powers of i follow a cycle of four. This means that $i^n = i^{n \bmod 4}$, where $n \bmod 4$ is the remainder when n is divided by 4. For example, $i^{58} = i^2 = -1$.

The imaginary numbers are numbers of the form bi, where b is a real number. The square root of any negative number is i times the square root of the positive of that number. Thus for example, $\sqrt{-9} = i\sqrt{9} = 3i$, $\sqrt{-12} = i\sqrt{12} = 2i\sqrt{3}$, and $\sqrt{-7} = i\sqrt{7}$.

EXERCISE

1. $i^{29} =$

   (A) 1
   (B) i
   (C) $-i$
   (D) -1
   (E) none of these

COMPLEX NUMBER ARITHMETIC

The complex numbers are formed by "attaching" imaginary numbers to real numbers using a plus sign (+). The standard form of a complex number is $a + bi$, where a and b are real. The number a is called the real part of the complex number, and the b number is called the imaginary part. If $b = 0$, then the complex number is just a real number. If $b \neq 0$, the complex number is called imaginary. If $a = 0$, bi is called a pure imaginary number. Examples of imaginary numbers are $2 + 3i$, $-\dfrac{3}{5} + 4i$, $6i$, $0.11 + (-0.45)i$, and $\pi - i\sqrt{5}$.

When the imaginary part of a complex number is a radical, write the i to the left in order to avoid ambiguity about whether i is under the radical.

Finding sums, differences, products, quotients, and reciprocals of complex numbers can be accomplished directly on your calculator. The imaginary unit i is 2nd decimal point. If you enter an expression with i in it, the calculator will do imaginary arithmetic in REAL mode. If the expression enetered does not include i but the output is imaginary, the calculator gives you the error message NONREAL ANS. For example, if you tried to calculate $\sqrt{-3}$ in REAL mode, you would get this error message. In $a + bi$ mode, $\sqrt{-3}$ would calculate as $1.732\cdots i$. You should use $a + bi$ mode exclusively for the Level 2 test. Although complex number arithmetic per se is not likely to be on a Level 2 test, an understanding of how it is done may be. A review of the main features of complex number arithmetic is provided in the next several paragraphs.

To add or subtract complex numbers, add or subtract their real and imaginary parts. For example,

$$(5 - 7i) + (2 + 4i) = 7 - 3i.$$

To multiply complex numbers, multiply like you would any two binomial expressions, using FOIL. Thus

$$(a + bi)(c + di) = ac + adi + bci + bdi^2 = (ac - bd) + (ad + bc)i.$$

The difference of the first and last terms makes the real part, and the sum of the outer and inner terms makes the imaginary part.

To find the quotient of two complex numbers, multiply the denominator and numerator by the conjugate of the denominator. Then simplify. For example,

$$\frac{2 - 7i}{3 + 5i} = \left(\frac{2 - 7i}{3 + 5i}\right) \cdot \left(\frac{3 - 5i}{3 - 5i}\right) = \frac{-29 - 31i}{9 + 25} = -\frac{29}{34} - \frac{31}{25}i.$$

EXERCISES

1. Write the product of $(2 + 3i)(4 - 5i)$ in standard form.

   (A) $-7 - 23i$
   (B) $-7 + 2i$
   (C) $23 - 7i$
   (D) $23 + 2i$
   (E) $23 - 2i$

2. Write $\dfrac{i}{2 - i}$ in standard form.

   (A) $-1 + \dfrac{1}{2}i$

   (B) $\dfrac{1}{5} - \dfrac{2}{5}i$

   (C) $-\dfrac{1}{5} + \dfrac{2}{5}i$

   (D) $-1 + 2i$
   (E) $-1 - 2i$

3. If $z = 8 - 2i$, $z^2 =$

    (A) $60 - 32i$
    (B) $64 + 4i$
    (C) $64 - 4i$
    (D) 60
    (E) 68

GRAPHING COMPLEX NUMBERS

A complex number can be represented graphically as rectangular coordinates, with the *x*-coordinate as the real part and the *y*-coordinate as the imaginary part. The modulus of a complex number is the square of its distance to the origin. The Pythagorean theorem tells us that this distance is $\sqrt{a^2 + b^2}$. The conjugate of the imaginary number $a + bi$ is $a - bi$, so the graphs of conjugates are reflections about the *y*-axis. Also, the product of an imaginary number and its conjugate is the modulus because $(a + bi)(a - bi) = a^2 - b^2i^2 = a^2 + b^2$.

EXERCISES

1. If *z* is the complex number shown in the figure, which of the following points could be *iz*?

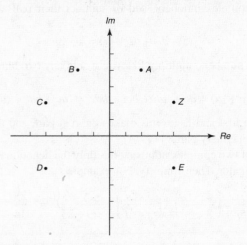

    (A) *A*
    (B) *B*
    (C) *C*
    (D) *D*
    (E) *E*

2. Which of the following is the modulus of $2 + i$?

    (A) $\sqrt{2}$
    (B) 2
    (C) $\sqrt{3}$
    (D) $\sqrt{5}$
    (E) 5

Answers and Explanations

In these solutions the following notation is used:

i: calculator unnecessary
a: calculator helpful or necessary
g: graphing calculator helpful or necessary

Imaginary Numbers

1. i **(B)** $i^{29} = i^1 = i$.

2. g **(C)** Simply enter the expression into the graphing calculator.

3. g **(A)** Simply enter the expression into the graphing calculator.

Complex Number Arithmetic

1. a **(D)** If you enter imaginary numbers into the calculator, it will do imaginary arithmetic without changing mode. The imaginary unit is 2nd decimal point. Enter the product, and read the solution $23 + 2i$.

Graphing Complex Numbers

1. i **(B)** $z = 4 + 2i$, so $iz = -2 + 4i$, which is point B.

2. i **(D)** The real and imaginary parts are 2 and 1, respectively, so the modulus is
$$\sqrt{2^2 + 1^2} = \sqrt{5}.$$

3.3 Matrices

ADDING, SUBTRACTING, AND SCALAR MULTIPLICATION

A matrix is a rectangular array of numbers. The size of a matrix is r by c, where r is the number of rows and c is the number of columns. The numbers in a matrix are called entries, and the entry in the ith row and jth column is named x_{ij}. If $r = 1$, the matrix is called a row matrix. If $c = 1$, the matrix is called a column matrix. If $r = c$, the matrix is called a square matrix. The numbers from the upper left corner to the bottom right corner of a square matrix form the main diagonal.

Scalar multiplication takes place when each number in a matrix is multiplied by a constant. If two matrices are the same size, they can be added or subtracted by adding or subtracting corresponding entries.

Example 1

$$3\begin{pmatrix} -2 & 3 \\ 1 & 5 \\ -4 & 3 \end{pmatrix} - 2\begin{pmatrix} 3 & -1 \\ 2 & 1 \\ -4 & 6 \end{pmatrix} = \begin{pmatrix} -12 & 11 \\ -1 & 13 \\ -4 & -3 \end{pmatrix}$$

1. $\begin{pmatrix} 1 & 3 \\ -2 & 4 \end{pmatrix} + \begin{pmatrix} 11 & 5 \\ -6 & 12 \end{pmatrix} = K \begin{pmatrix} 3 & 2 \\ J & M \end{pmatrix}$. Find the value of $K + J + M$.

(A) 2
(B) 4
(C) 6
(D) 7
(E) 8

[handwritten: K = 4, J = -2, M = 4]

MATRIX MULTIPLICATION

Matrix multiplication takes place when two matrices, A and B, are multiplied to form a new matrix, AB. Matrix multiplication is possible only under certain conditions. Suppose A is r_1 by c_1 and B is r_2 by c_2. If $c_1 = r_2$, then AB is defined and has size r_1 by c_2. The entry x_{ij} of AB is the ith row of A times the jth column of B. If A and B are square matrices, BA is also defined but not generally equal to AB.

Example 1

Evaluate $AB = \begin{pmatrix} 3 & -1 \\ 3 & 5 \\ -2 & 1 \end{pmatrix} \begin{pmatrix} 1 & -2 \\ 5 & -3 \end{pmatrix}$. *A* is 3 by 2, and *B* is 2 by 2.

The product matrix is 3 by 2, with entries x_{ij} calculated as follows:

$x_{11} = (3)(1) + (-1)(5) = -2$, the "product" of row 1 of A and column 1 of B

$x_{12} = (3)(-2) + (-1)(-3) = -3$, the "product" of row 1 of A and column 2 of B

$x_{21} = (3)(1) + (5)(5) = 28$, the "product" of row 2 of A and column 1 of B

$x_{22} = (3)(-2) + (5)(-3) = -21$, the "product" of row 2 of A and column 2 of B

$x_{31} = (-2)(1) + (1)(5) = 3$, the "product" of row 3 of A and column 1 of B

$x_{32} = (-2)(-2) + (1)(-3) = 1$, the "product" of row 3 of A and column 2 of B

Therefore, $AB = \begin{pmatrix} 3 & -1 \\ 3 & 5 \\ -2 & 1 \end{pmatrix} \begin{pmatrix} 1 & -2 \\ 5 & -3 \end{pmatrix} = \begin{pmatrix} -2 & -3 \\ 28 & -21 \\ 3 & 1 \end{pmatrix}$. Note that BA is not defined.

Matrix calculations can be done on a graphing calculator. To define a matrix, enter 2nd MATRIX, highlight and enter EDIT, enter the number of rows followed by the number of columns, and finally enter the entries. The figure below shows the result of these steps for matrices A and B of Example 1.

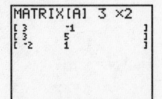
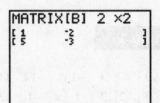

To find the product, enter 2nd MATRIX/NAMES/[A], which returns [A] to the home screen. Also enter 2nd MATRIX/NAMES/[B], which returns [B] to the home screen. Hit ENTER again to get the product.

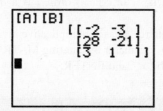

EXERCISE

1. The chart below shows the number of small and large packages of a certain brand of cereal that were bought over a three-day period. The price of a small box of this brand is $2.99, and the price of a large box is $3.99. Which of the following matrix expressions represents the income, in dollars, received from the sale of cereal each of the three days?

	Day 1	Day 2	Day 3
Large	75	82	57
Small	43	36	50

(A) $\begin{pmatrix} 75 & 82 & 57 \\ 43 & 36 & 50 \end{pmatrix}(2.99 \quad 3.99)$

(B) $\begin{pmatrix} 75 & 43 \\ 82 & 36 \\ 57 & 50 \end{pmatrix}\begin{pmatrix} 3.99 \\ 2.99 \end{pmatrix}$

(C) $\begin{pmatrix} 75 & 82 & 57 \\ 43 & 36 & 50 \end{pmatrix}\begin{pmatrix} 2.99 \\ 3.99 \end{pmatrix}$

(D) $\begin{pmatrix} 2.99 \\ 3.99 \end{pmatrix}\begin{pmatrix} 75 & 43 \\ 82 & 36 \\ 57 & 50 \end{pmatrix}$

(E) $2.99\begin{pmatrix} 75 & 82 & 57 \\ 43 & 36 & 50 \end{pmatrix} + 3.99\begin{pmatrix} 75 & 82 & 57 \\ 43 & 36 & 50 \end{pmatrix}$

DETERMINANTS AND INVERSES OF SQUARE MATRICES

The determinant of an n by n square matrix is a number. The determinant of the 2 by 2 matrix $\begin{pmatrix} a & b \\ c & d \end{pmatrix}$ is $ad - bc$. For larger square matrices, use the graphing calculator to calculate the determinant (2nd MATRIX/MATH/det). A matrix whose determinant is zero is called singular. If the determinant is not zero, the matrix is nonsingular.

The product of square n by n matrices is a square n by n matrix. An identity matrix I is a square matrix consisting of 1's down the main diagonal and 0's elsewhere. The product of n by n square matrices I and A is A. In other words, I is a multiplicative identity for matrix multiplication.

A nonsingular square n by n matrix A has a multiplicative inverse, A^{-1}, where $A^{-1}A = AA^{-1} = I$. A^{-1} can be found on the graphing calculator by entering MATRIX/NAMES/A, which will return A to the home screen, followed by x^{-1} and ENTER.

EXERCISE

1. Find all values of x for which $\begin{vmatrix} 2 & -1 & 4 \\ 3 & 0 & 5 \\ 4 & 1 & 6 \end{vmatrix} = \begin{vmatrix} x & 4 \\ 5 & x \end{vmatrix}$.

   (A) ±3.78
   (B) ±4.47
   (C) ±5.12
   (D) ±6.19
   (E) ±6.97

SOLVING SYSTEMS OF EQUATIONS

An important application of matrices is writing and solving systems of equations in matrix form.

Example 1

Solve the system
$$\begin{aligned} x - y + 2z &= -3 \\ 2x + y - z &= 0 \\ -x + 2y - 3z &= 7 \end{aligned}$$

This system can be written as $AV = B$, where $A = \begin{pmatrix} 1 & -1 & 2 \\ 2 & 1 & -1 \\ -1 & 2 & -3 \end{pmatrix}$ is the matrix of the

coefficients, $V = \begin{pmatrix} x \\ y \\ z \end{pmatrix}$ is the matrix of the variables, and $B = \begin{pmatrix} -3 \\ 0 \\ 7 \end{pmatrix}$ is the column matrix

representing the right side of the system. Multiplying both sides of this equation on the left by A^{-1} yields $A^{-1}AV = A^{-1}B$, which reduces to $V = A^{-1}B$. If given a system of equations, enter the matrix of coefficients into a matrix (A), the column matrix of the right side into a second matrix (B), and find the product $A^{-1}B$ to display the solution.

EXERCISE

1. Find $\begin{pmatrix} x \\ y \end{pmatrix}$ if $\begin{pmatrix} 3 & 2 \\ -1 & 4 \end{pmatrix}\begin{pmatrix} x \\ y \end{pmatrix} = \begin{pmatrix} -5 \\ 4 \end{pmatrix}$.

(A) $(-2 \quad 0.5)$

(B) $\begin{pmatrix} -5/6 \\ 1 \end{pmatrix}$

(C) $(-1 \quad 3/4)$

(D) $\begin{pmatrix} -2 \\ 1/2 \end{pmatrix}$

(E) $(-5 \quad -4/5)$

Answers and Explanations

In these solutions the following notation is used:

    i: calculator unnecessary
    a: calculator helpful or necessary
    g: graphing calculator helpful or necessary

Adding, Subtracting, and Scalar Multiplication

1. i   **(C)** First find the sum of the matrices to the left of the equals sign: $\begin{pmatrix} 12 & 8 \\ -8 & 16 \end{pmatrix}$.

Since the first row of the matrix to the right of the equals sign is $(3 \quad 2)$, K must be 4. Since $(J \quad M)$ is the bottom row, $J = -2$ and $M = 4$. Therefore, $K + J + M = 6$.

Matrix Multiplication

1. i   **(B)** Matrix multiplication is row by column. Since the answer must be a 3 by 1 matrix, the only possible answer choice is B.

Determinants and Inverses of Square Matrices

1. a   **(B)** Enter the 3 by 3 matrix on the left side of the equation into your graphing calculator and evaluate its determinant (zero). The determinant on the right side of the equation is $x^2 - 20$. Therefore $x = \pm\sqrt{20} \approx \pm 4.47$.

Solving Systems of Equations

1. a   **(D)** This is the matrix form $AX = B$ of a system of equations. Multiply both sides of

the equation by A^{-1} on the left to get the solution $X = \begin{pmatrix} x \\ y \end{pmatrix} = A^{-1}B$. Enter the 2 by 2

matrix, A, and the 2 by 1 matrix, B, into your graphing calculator. Return to the home

screen and enter $A^{-1}B = \begin{pmatrix} -2 \\ 0.5 \end{pmatrix}$.

3.4 Sequences and Series

RECURSIVE SEQUENCES

A *sequence* is a function with a domain consisting of the natural numbers. A *series* is the sum of the terms of a sequence.

Example 1

Give an example of
(A) an infinite sequence of numbers,
(B) a finite sequence of numbers,
(C) an infinite series of numbers.

SOLUTIONS

(A) $\dfrac{1}{2}, \dfrac{1}{3}, \dfrac{1}{4}, \dfrac{1}{5}, \dots, \dfrac{1}{n+1}, \dots$, is an *infinite* sequence of numbers with

$$t_1 = \frac{1}{2}, \ t_2 = \frac{1}{3}, \ t_3 = \frac{1}{4}, \ t_4 = \frac{1}{5}, \text{ and } t_n = \frac{1}{n+1}.$$

(B) $2, 4, 6, \dots, 20$ is a *finite* sequence of numbers with $t_1 = 2, t_2 = 4, t_3 = 6, t_{10} = 20$.

(C) $\dfrac{1}{2} + \dfrac{1}{4} + \dfrac{1}{8} + \dfrac{1}{16} + \dots + \dfrac{1}{2^n} + \dots$ is an infinite series of numbers.

Example 2

If $t_n = \dfrac{2n}{n+1}$, find the first five terms of the sequence.

When 1, 2, 3, 4, and 5 are substituted for n, $t_1 = \dfrac{2}{2} = 1, t_2 = \dfrac{4}{3}, \ t_3 = \dfrac{6}{4} = \dfrac{3}{2}, t_4 = \dfrac{8}{5}$, and

$t_5 = \dfrac{10}{6} = \dfrac{5}{3}$.

The first five terms are $1, \ \dfrac{4}{3}, \dfrac{3}{2}, \dfrac{8}{5}, \dfrac{5}{3}$.

Example 3

If $a_1 = 2$ and $a_n = \dfrac{a_{n-1}}{2}$, find the first five terms of the sequence.

Since every term is expressed with respect to the immediately preceding term, this is called a *recursion formula*, and the resulting sequence is called a recursive sequence.

$$a_1 = 2, \, a_2 = \frac{a_1}{2} = \frac{2}{2} = 1,$$

$$a_3 = \frac{a_2}{2} = \frac{1}{2}, \, a_4 = \frac{a_3}{2} = \frac{\frac{1}{2}}{2} = \frac{1}{4},$$

$$a_5 = \frac{a_4}{2} = \frac{\frac{1}{4}}{2} = \frac{1}{8}.$$

Therefore, the first five terms are $2, 1, \dfrac{1}{2}, \dfrac{1}{4}, \dfrac{1}{8}$.

Example 4

If $a_1 = 3$ and $a_n = 2a_{n-1} + 5$, find a_4.

Put $3(a_1)$ into your graphing calculator, and press ENTER. Then multiply by 2 and add 5. Hit ENTER 3 more times to get $a_4 = 59$.

Example 5

If $a_1 = 1$, $a_2 = 1$, and $a_n = a_{n-1} + a_{n-2}$ for $n \geq 3$, find the first 7 terms of the sequence.

The recursive formula indicates that each term is the sum of the two terms before it. Therefore, the first seven terms of this sequence are 1, 1, 2, 3, 5, 8, 13.

A series can be abbreviated by using the Greek letter sigma, Σ, to represent the summation of several terms.

Example 6

(A) Express the series $2 + 4 + 6 + \cdots + 20$ in sigma notation.

(B) Evaluate $\displaystyle\sum_{k=0}^{5} k^2$

SOLUTIONS

(A) The series $2 + 4 + 6 + \cdots + 20 = \displaystyle\sum_{i=1}^{10} 2i = 110$.

(B) $\displaystyle\sum_{k=0}^{5} k^2 = 0^2 + 1^2 + 2^2 + 3^2 + 4^2 + 5^2 = 0 + 1 + 4 + 9 + 16 + 25 = 55$.

ARITHMETIC SEQUENCES

One of the most common sequences studied at this level is an *arithmetic sequence* (or *arithmetic progression*). Each term differs from the preceding term by a common difference. The first n terms of an arithmetic sequence can be denoted by

$$t_1, t_1 + d, t_1 + 2d, t_1 + 3d, \ldots, t_1 + (n-1)d$$

where d is the common difference and $t_n = t_1 + (n-1)d$. The sum of n terms of the series constructed from an arithmetic sequence is given by the formula

$$S_n = \frac{n}{2}(t_1 + t_n) \quad \text{or} \quad S_n = \frac{n}{2}[2t_1 + (n-1)d]$$

If there is one term falling between two given terms of an arithmetic sequence, it is called their arithmetic mean.

Example 1

(A) **Find the 28th term of the arithmetic sequence 2, 5, 8, ...**
(B) **Express the sum of 28 terms of the series of this sequence using sigma notation.**
(C) **Find the sum of the first 28 terms of the series.**

SOLUTIONS

(A) $t_n = t_1 + (n-1)d$

$t_1 = 2, d = 3, n = 28$

$t_{28} = 2 + 27 \cdot 3 = 83$

(B) $\displaystyle\sum_{k=0}^{27}(3k+2)$ or $\displaystyle\sum_{j=1}^{28}(3j-1)$

(C) $S_n = \frac{n}{2}(t_1 + t_n)$

$S_{28} = \frac{28}{2}(2+83) = 14 \cdot 85 = 1190$

Example 2

If $t_8 = 4$ and $t_{12} = -2$, **find the first three terms of the arithmetic sequence.**

$$t_n = t_1 + (n-1)d$$
$$4 = t_1 + 7d$$
$$-2 = t_1 = 11d$$

To solve these two equations for d, subtract the second equation from the third.

$$-6 = 4d$$
$$d = -\frac{3}{2}$$

Substituting in the first equation gives $4 = t_1 + 7\left(-\dfrac{3}{2}\right)$. Thus,

$$t_1 = 4 + \frac{21}{2} = \frac{29}{2}$$

$$t_2 = \frac{29}{2} + \left(-\frac{3}{2}\right) = \frac{26}{2} = 13$$

$$t_3 = \frac{29}{2} + 2\left(-\frac{3}{2}\right) = \frac{23}{2}$$

The first three terms are $\dfrac{29}{2}, 13, \dfrac{23}{2}$.

Example 3

In an arithmetic series, if $S_n = 3n^2 + 2n$, find the first three terms.

When $n = 1$, $S_1 = t_1$. Therefore, $t_1 = 3(1)^2 + 2 \cdot 1 = 5$.

$$
\begin{aligned}
S_2 = t_1 + t_2 &= 3(2)^2 + 2 \cdot 2 = 16 \\
5 + t_2 &= 16 \\
t_2 &= 11
\end{aligned}
$$

Therefore, $d = 6$, which leads to a third term of 17. Thus, the first three terms are 5, 11, 17.

GEOMETRIC SEQUENCES

Another common type of sequence studied at this level is a geometric sequence (or geometric progression). In a geometric sequence the ratio of any two successive terms is a constant r called the constant ratio. The first n terms of a geometric sequence can be denoted by

$$t_1, t_1 r, t_1 r^2, t_1 r^3, \cdots, t_1 r^{n-1} = t_n$$

The sum of the first n terms of a geometric series is given by the formula

$$Sn = \frac{t_1\left(1 - r^n\right)}{1 - r}.$$

If there is one term falling between two given terms of a geometric sequence it is called their geometric mean.

Example 1

(A) Find the seventh term of the geometric sequence 1, 2, 4, . . . , and
(B) the sum of the first seven terms.

(A) $r = \dfrac{t_2}{t_1} = \dfrac{2}{1} = 2; t_7 = t_1 r^{7-1}; t_7 = 1 \cdot 2^6 = 64$

(B) $S_7 = \dfrac{1\left(1 - 2^7\right)}{1 - 2} = \dfrac{1 - 128}{-1} = 127$

Example 2

The first term of a geometric sequence is 64, and the common ratio is $\frac{1}{4}$. For what value of n is $t_n = \frac{1}{4}$?

$$\frac{1}{4} = 64\left(\frac{1}{4}\right)^{n-1}$$

$$\left(\frac{1}{4}\right)^{2-n} = 64 = 4^3$$

$$4^{n-2} = 4^3$$

$$n = 5$$

SERIES

In a geometric sequence, if $|r| < 1$, the sum of the series approaches a limit as n approaches infinity. In the formula $S_n = \dfrac{t_1\left(1 - r^n\right)}{1 - r}$, if $|r| < 1$, the term $r^n \to 0$ as $n \to \infty$. Therefore, as long as $|r| < 1$, $\displaystyle\lim_{n \to \infty} S_n = \dfrac{t_1}{1 - r}$, or $S = \dfrac{t_1}{1 - r}$.

Example 1

Evaluate (A) $\displaystyle\lim_{n \to \infty} \sum_{k=1}^{n} \frac{1}{2^k}$ and

(B) $\displaystyle\sum_{j=0}^{\infty} (-3)^{-j}$

Both problems ask the same question: Find the sum of an infinite geometric series.

(A) When the first few terms, $\dfrac{1}{2} + \dfrac{1}{4} + \dfrac{1}{8} + \cdots$, are listed, it can be seen that $t_1 = \dfrac{1}{2}$ and the common ratio $r = \dfrac{1}{2}$. Therefore,

$$S = \frac{\dfrac{1}{2}}{1 - \dfrac{1}{2}} = 1.$$

(B) When the first few terms, $\dfrac{1}{1} - \dfrac{1}{3} + \dfrac{1}{9} - \cdots$, are listed, it can be seen that $t_1 = 1$ and the common ratio $r = -\dfrac{1}{3}$. Therefore,

$$S = \frac{1}{1 - \left(-\dfrac{1}{3}\right)} = \frac{1}{\dfrac{4}{3}} = \frac{3}{4}.$$

Example 2

Find the exact value of the repeating decimal 0.4545 . . . .

This can be represented by a geometric series, $0.45 + 0.0045 + 0.000045 + \cdots$, with $t_1 = 0.45$ and $r = 0.01$.

Since $|r| < 1$,

$$S = \frac{0.45}{1-0.01} = \frac{0.45}{0.99} = \frac{45}{99} = \frac{5}{11}.$$

Example 3

Given the sequence 2, x, y, 9. If the first three terms form an arithmetic sequence and the last three terms form a geometric sequence, find x and y.

From the arithmetic sequence $\begin{cases} x = 2+d \\ y = 2+2d \end{cases}$, substitute to eliminate d.

$$y = 2 + 2(x-2)$$
$$y = 2 + 2x - 4$$
$$*y = 2x - 2$$

From the geometric sequence $\begin{cases} 9 = yr \\ y = xr \end{cases}$, substitute to eliminate r.

$$9 = y \cdot \frac{y}{x}$$
$$* 9x = y^2$$

Use the two equations with the * to eliminate y:

$$9x = (2x-2)^2$$
$$9x = 4x^2 - 8x + 4$$
$$4x^2 - 17x + 4 = 0$$
$$(4x-1)(x-4) = 0$$
$$4x - 1 = 0 \quad \text{or} \quad x - 4 = 0$$

Thus, $x = \frac{1}{4}$ or 4.

Substitute in $y = 2x - 2$:

$$\text{if } x = \frac{1}{4}, \ y = -\frac{3}{2}$$
$$\text{if } x = 4, \ y = 6.$$

EXERCISES

1. If $a_1 = 3$ and $a_n = n + a_{n-1}$, the sum of the first five terms is

   (A) 17
   (B) 30
   (C) 42
   (D) 45
   (E) 68

2. If $a_1 = 5$ and $a_n = 1 + \sqrt{a_{n-1}}$, find a_3.

   (A) 2.623
   (B) 2.635
   (C) 2.673
   (D) 2.799
   (E) 3.323

3. If the repeating decimal $0.237\overline{37} \ldots$ is written as a fraction in lowest terms, the sum of the numerator and denominator is

   (A) 16
   (B) 47
   (C) 245
   (D) 334
   (E) 1237

4. The first three terms of a geometric sequence are $\sqrt[4]{3}, \sqrt[8]{3}$, 1. The fourth term is

   (A) $\sqrt[32]{3}$

   (B) $\sqrt[16]{3}$

   (C) $\dfrac{1}{\sqrt[16]{3}}$

   (D) $\dfrac{1}{\sqrt[8]{3}}$

   (E) $\dfrac{1}{\sqrt[4]{3}}$

5. By how much does the arithmetic mean between 1 and 25 exceed the positive geometric mean between 1 and 25?

   (A) 5
   (B) about 7.1
   (C) 8
   (D) 12.9
   (E) 18

6. In a geometric series $S = \frac{2}{3}$ and $t_1 = \frac{2}{7}$. What is r?

(A) $\frac{2}{3}$

(B) $-\frac{4}{7}$

(C) $\frac{2}{7}$

(D) $\frac{4}{7}$

(E) $-\frac{2}{7}$

Answers and Explanations

In these solutions the following notation is used:

    i: calculator unnecessary
    a: calculator helpful or necessary
    g: graphing calculator helpful or necessary

1. a   **(D)** $a_2 = 5$, $a_3 = 8$, $a_4 = 12$, $a_5 = 17$. Therefore, $S_5 = 45$.

2. i   **(B)** Press 5 ENTER into your graphing calculator. Then enter $\sqrt{(\text{Ans})+1}$ and press ENTER twice more to get a_3.

3. a   **(C)** The decimal $0.23\overline{37} = 0.2 + (0.037 + 0.00037 + 0.0000037 + \cdots)$, which is 0.2 + an infinite geometric series with a common ratio of 0.01.

$$S_n = 0.2 + \frac{0.037}{0.99} = \frac{2}{10} + \frac{37}{990} = \frac{235}{990} = \frac{47}{198}.$$

The sum of the numerator and the denominator is 245.

4. i   **(D)** Terms are $3^{1/4}$, $3^{1/8}$, 1. Common ratio $= 3^{-1/8}$. Therefore, the fourth term is

$1 \cdot 3^{-1/8} = 3^{-1/8}$ or $\dfrac{1}{\sqrt[8]{3}}$.

5. i   **(C)** Arithmetic mean $= \dfrac{1+25}{2} = 13$. Geometric mean $= \sqrt{1 \cdot 25} = 5$. The difference is 8.

6. i   **(D)** $\dfrac{2}{3} = \dfrac{\frac{2}{7}}{1-r}$. $2 - 2r = \dfrac{6}{7}$. $14 - 14r = 6$. Therefore, $r = \dfrac{4}{7}$.

3.5 Vectors

VECTORS

A vector in a plane is defined to be an ordered pair of real numbers. A vector in space is defined as an ordered triple of real numbers. On a coordinate system, a vector is usually represented by an arrow whose initial point is the origin and whose terminal point is at the ordered pair (or triple) that named the vector. Vector quantities always have a magnitude or *norm* (the length of the arrow) and direction (the angle the arrow makes with the positive *x*-axis). Vectors are often used to represent motion or force.

All properties of two-dimensional vectors can be extended to three-dimensional vectors. We will express the properties in terms of two-dimensional vectors for convenience. If vector $\vec{V}$ is designated by (v_1, v_2) and vector $\vec{U}$ is designated by (u_1, u_2), vector $\overrightarrow{U+V}$ is designated by $(u_1 + v_1, u_2 + v_2)$ and called the *resultant* of $\vec{U}$ and $\vec{V}$. Vector $-\vec{V}$ has the same magnitude as $\vec{V}$ but has a direction opposite that of $\vec{V}$.

On the plane, every vector $\vec{V}$ can be expressed in terms of any other two unit (magnitude 1) vectors parallel to the *x*- and *y*-axes. If vector $\vec{i} = (1,0)$ and vector $\vec{j} = (0,1)$, any vector $\vec{V} = ai + bj$, where *a* and *b* are real numbers. A unit vector parallel to $\vec{V}$ can be determined by dividing $\vec{V}$ by its norm, denoted by $\| \vec{V} \|$ and equal to $\sqrt{a^2 + b^2}$

It is possible to determine algebraically whether two vectors are perpendicular by defining the *dot product* or *inner product* of two vectors, $\vec{V}(v_1, v_2)$ and $\vec{U}(u_1, u_2)$.

$$\vec{V} \cdot \vec{U} = v_1 u_1 + v_2 u_2$$

Notice that the dot product of two vectors is a *real number*, not a vector. Two vectors, $\vec{V}$ and $\vec{U}$, are perpendicular if and only if $\vec{V} \cdot \vec{U} = 0$.

Example 1

Let vector $\vec{V} = (2,3)$ and vector $\vec{U} = (6,-4)$.

(A) What is the resultant of $\vec{U}$ and $\vec{V}$?

(B) What is the norm of $\vec{U}$?

(C) Express $\vec{V}$ in terms of $\vec{i}$ and $\vec{j}$.

(D) Are $\vec{U}$ and $\vec{V}$ perpendicular?

SOLUTIONS

(A) The resultant, $\overrightarrow{U+V}$, equals $(6 + 2, -4 + 3) = (8, -1)$.

(B) The norm of $\vec{U}$, $\|\vec{U}\| = \sqrt{36+16} = \sqrt{52} = 2\sqrt{13}$.

(C) $\vec{V} = 2\vec{i} + 3\vec{j}$. To verify this, use the definitions of $\vec{i}$ and $\vec{j}$. $\vec{V} = 2(1,0) + 3(0,1) =$

$(2,0) + (0,3) = (2,3) = \vec{V}$.

(D) $\vec{U} \cdot \vec{V} = 6 \cdot 2 + (-4) \cdot 3 = 12 - 12 = 0$. Therefore, $\vec{U}$ and $\vec{V}$ are perpendicular because the dot product is equal to zero.

Example 2

If $\vec{U} = (-1,4)$ and the resultant of $\vec{U}$ and $\vec{V}$ is $(4,5)$, find $\vec{V}$.

Let $\vec{V} = (v_1, v_2)$. The resultant $\overrightarrow{U+V} = (-1,4) + (v_1, v_2) = (4,5)$. Therefore, $(-1 + v_1, 4 + v_2) =$

$(4,5)$, which implies that $-1 + v_1 = 4$ and $4 + v_2 = 5$. Thus, $v_1 = 5$ and $v_2 = 1$. $\vec{V} = (5,1)$.

EXERCISES

1. Suppose $\vec{x} = (-3,-1)$, $\vec{y} = (-1,4)$. Find the magnitude of $\vec{x} + \vec{y}$.

   (A) 2
   (B) 3
   (C) 4
   (D) 5
   (E) 6

2. If $\vec{V} = 2\vec{i} + 3\vec{j}$ and $\vec{U} = \vec{i} - 5\vec{j}$, the resultant vector of $2\vec{U} + 3\vec{V}$ equals

   (A) $3\vec{i} - 2\vec{j}$

   (B) $5\vec{i} + \vec{j}$

   (C) $7\vec{i} - 9\vec{j}$

   (D) $8\vec{i} - \vec{j}$

   (E) $2\vec{i} + 3\vec{j}$

3.   A unit vector perpendicular to vector $\vec{V} = (3, -4)$ is

(A) $(4, 3)$

(B) $\left(\dfrac{3}{5}, \dfrac{4}{5}\right)$

(C) $\left(-\dfrac{3}{5}, -\dfrac{4}{5}\right)$

(D) $\left(-\dfrac{4}{5}, -\dfrac{3}{5}\right)$

(E) $\left(-\dfrac{4}{5}, \dfrac{3}{5}\right)$

Answers and Explanations

In these solutions the following notation is used:

　　i: calculator unnecessary
　　a: calculator helpful or necessary
　　g: graphing calculator helpful or necessary

1. i   **(D)** Add the components to get $\vec{x} + \vec{y} = (-4, 3)$. The magnitude is $\sqrt{(-4)^2 + 3^2} = 5$.

2. i   **(D)** $2\vec{U} = 2\vec{i} - 10\vec{j}$ and $3\vec{V} = 6\vec{i} + 9\vec{j}$, so $2\vec{U} + 3\vec{V} = 8\vec{i} - \vec{j}$.

3. i   **(D)** All answer choices except A are unit vectors. Backsolve to find that the only one having a zero dot product with $(3, -4)$ is $\left(-\dfrac{4}{5}, -\dfrac{3}{5}\right)$.

Data Analysis, Statistics, and Probability

• Data Analysis and Statistics	• Probability

4.1 Data Analysis and Statistics

USING STATISTICS TO ANALYZE DATA

The word *data* means a set of numbers that refer to a variable. Examples are test scores, heights, salaries, number of cigarettes smoked each day, and number of deaths in airplane crashes. One type of data analysis groups the numbers into value ranges and counts how many numbers are in each group. These analyses result in distributions of the data. Data distributions are often presented as visual displays such as histograms and boxplots.

Data can be summarized by measuring their center and spread. The mean of a data set is the sum of the values divided by the number of values. Standard deviation is the measure of spread that accompanies the mean. Loosely speaking, the standard deviation of a data set is the average deviation of individual values from their mean. If $x_1, x_2, \ldots, x_n$ are the data values, the mean is

$\bar{x} = \frac{1}{n} \sum x_i$, and the standard deviation is $s = \sqrt{\frac{1}{n-1} \sum \left(x_i - \bar{x} \right)^2}$. The symbol $\sum$ (sigma) indicates

the sum of the quantities to its immediate right. The square of the standard deviation s^2 is called the variance of the data set. Generally speaking, using mean and standard deviation to summarize a data set is appropriate only when the set is unimodal and symmetric, such as the one shown in the figure in Example 1 on the next page.

The median of a data set is the middle value when the values are ordered from smallest to largest. The median is close in value to the mean when a data set is unimodal and symmetric. A data set whose distribution has a long tail on one side, such as the one shown below, is called skew. The spread of a skew distribution is best measured by the five-number summary: minimum, maximum, median, and two quartiles (defined as the medians of the two halves determined by the median).

The difference between the maximum and the minimum is called the range of the data set. The difference between the higher quartile (Q_3) and lower quartile (Q_1) is called the interquartile range (IQR). Since the median of a data set is based on *numbers* of values in certain intervals and not on *actual* values in the data set, it, along with the five-number summary, should be used to summarize skew data.

The mode of a data set is the value that occurs most frequently. For example, in the data set 1, 6, 3, 6, 4, 6, 2, 0, 6, 6, the mode is 6 because it appears 5 times—more often than any other value. If no value appears more frequently than the others, the data set has no mode. If two values appear with equal frequency and more often than other values, the data set is said to be bimodal.

The mode is more useful when working with categorical variables, such as hair color, gender, breed of dog, and so on. In these cases, the mode is the category that has the largest frequency.

Example 1

The final exam in a statistics course resulted in the following 40 grades (ordered from smallest to largest):

51	53	57	59	64	65	65	66	68	69
69	71	71	71	73	73	74	75	75	75
76	77	77	78	79	80	81	82	82	83
83	84	85	88	90	91	95	96	98	98

Construct a histogram using these data, and comment on the shape of the histogram.

Enter the data into a list (L1) in the graphing calculator. Press 2ndY= and select Plot 1. Select On, the histogram logo, Xlist L1, and Freq 1. Set the WINDOW values to Xmin = 50, Xmax = 100, Xscl = 10, Ymin = –5, Ymax = 20, and Yscl = 5. Then press TRACE to get the following screen.

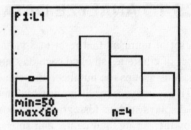

The distribution of these test scores is unimodal and symmetric.

Example 2

Find the mean and standard deviation of the test scores in Example 1.

Press 2nd/LIST/MATH/3/ENTER on the graphing calculator, which returns mean (to the home screen). Press 2nd/L1/ENTER to display the mean of 76.175. Next press 2nd/LIST/MATH/7/ENTER, which returns stdDev(to the home screen). Again, press 2nd/L1/ENTER, this time displaying the standard deviation of 11.564 .

Example 3

The table below shows the heights (inches) of 130 choir members.

Height	Count	Height	Count
60	2	69	5
61	6	70	11
62	9	71	8
63	7	72	9
64	5	73	4
65	20	74	2
66	18	75	4
67	7	76	1
68	12		

Construct a boxplot for the five-number summary.

Enter the heights into one list (L2) and the counts into another (L3). Press 2ndY= and select a plot. Turn the plot on, press the right arrow to the boxplot logo, and enter L2 for Xlist and L3 for Freq. The command "Mark" selects the symbol to indicate outliers.

Press ZOOM/9 for the calculator to select a scale for the boxplot. Pressing TRACE brings the flashing cursor to the screen to highlight the five numbers. A boxplot for the choir data is shown below.

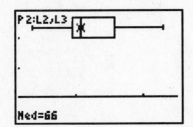

The five-number summary is minimum = 60, Q1 = 65, median = 66, Q3 = 70, and maximum = 76.

EXERCISES

1. Last week, police ticketed 13 men traveling 18 miles per hour over the speed limit and 8 women traveling 14 miles per hour over the speed limit. What was the mean speed over the limit of all 21 drivers?

   (A) 16 miles per hour
   (B) 16.5 miles per hour
   (C) 17 miles per hour
   (D) none of these
   (E) cannot be determined

2. If the range of a set of integers is 2 and the mean is 50, which of the following statements must be true?

   I. The mode is 50
   II. The median is 50 ✓
   III. There are exactly three data values

   (A) only I
   (B) only II
   (C) only III
   (D) I and II
   (E) I, II, and III

3. What is the median of the frequency distribution shown below?

Data Value	Frequency
0	1
1	3
2	7
3	15
4	10
5	7
6	3
7	3

   (A) 2
   (B) 3
   (C) 4
   (D) 5
   (E) Cannot be determined

REGRESSION

Regression is used to analyze the relationship between two variables. This is accomplished by an equation that expresses one of the variables (response/dependent/y-variable) as a function of the other (explanatory/independent/x-variable). A scatterplot is a graphical display of two data sets as points on a rectangular coordinate system. An examination of the scatterplot indicates the type of function that best models the data. The Level 2 test considers linear, quadratic, and exponential functions to model the relationship between two variables. Generally, a question would state which type of model to use with given data sets.

With the data sets in two lists, the numerical values needed for a function are determined by the graphing calculator. These commands are in STAT/CALC menu. The least squares regression equation can be stored in the Y= menu by selecting from the VARS/Y-VARS/FUNCTION menu and including that selection after naming the lists for the regression.

Example 4

The decennial population of Center City for the past five decades is shown in the table below. Use exponential regression to estimate the 1965 population.

Population of Center City

Year	Population
1950	48,000
1960	72,000
1970	95,000
1980	123,000
1990	165,000

Transform the years to "number of years after 1950" and enter these values into L4. Then enter the populations in thousands Set up the scatterplot by pressing 2ndY= and selecting a plot (Plot 1). Turn the plot on, select the scatterplot logo, and enter the list names. Then press STAT/CALC/ExpReg L4,L5,Y1. This will store the regression equation in Y1. The resulting command is shown in the left screen below. Press ENTER to display the values for the equation. These are shown in the right screen below.

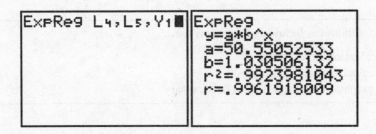

Press ZOOM/9 to view the scatterplot and exponential curve. Press 2nd/CALC/value and enter 15, representing 1965. The cursor moves to the point on the regression curve where $x = 15$ and displays both x and y at the bottom of the screen, as shown below.

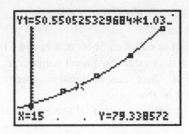

The 1965 population was about 79,300.

EXERCISE
---

1. Jack recorded the amount of time he studied the night before each of 4 history quizzes and the score he got on each quiz. The data are in the table below.

Score	Time (min.)
86	45
70	15
90	40
78	35

Use linear regression to estimate the score Jack would get if he studied for 20 minutes.

(A) 71
(B) 72
(C) 73
(D) 74
(E) 75

Answers and Explanations

In these solutions the following notation is used:

    i: calculator unnecessary
    a: calculator helpful or necessary
    g: graphing calculator helpful or necessary

Using Statistics to Analyze Data

1. a   **(B)** There are 13 eighteens and 8 fourteens, so the total over the speed limit is 346. Divide this by the 21 people to get 16.5.

2. i   **(B)** Since the data values are integers, the range is 2, and the mean is 50, the possible data values are 49, 50, and 51.
   I. The set could consist of equal numbers of 49s and 51s and have a mean of 50 without 50 even being a data value. So I need not be true.
   II. Since the mean is 50, there must be equal numbers of 49s and 51s, so 50 is also the median. II must be true.
   III. Explanations in 1 and 2 imply that III need not be true.

3. i   **(B)** There are 49 data values altogether, so the median is the 25th largest. Adding the frequencies up to 25 puts the 25th number at 3.

Relationships Between Variables

1. a   **(C)** Enter the data in two lists (study times in L1 and test scores in L2). Enter STAT/CALC/8, and enter VARS/YVARS/Function/Y_1, followed by ENTER. This produces estimates of the slope (b) and y-intercept (a) of the regression line $a + bx$. Enter this expression into Y_1. Enter $Y_1(2)$ to get the score of 72.

4.2 Probability

The probability of an event happening is a number defined to be the number of ways the event can happen successfully divided by the total number of ways the event can happen.

Example 1

What is the probability of getting a head when a coin is flipped?

A coin can fall in one of two ways, heads or tails, and each is equally likely.

$$P(\text{head}) = \frac{\text{number of ways a head can come up}}{\text{total number of ways the coin can fall}}$$

$$= \frac{1}{2}.$$

Example 2

What is the probability of getting a 3 when one die is thrown?

A die can fall with any one of six different numbers showing, and there is only one way a 3 can show.

$$P(3) = \frac{\text{number of ways a 3 can come up}}{\text{total number of ways the die can fall}} = \frac{1}{6}.$$

Example 3

What is the probability of getting a sum of 7 when two dice are thrown?

Since it is not obvious how many different throws will produce a sum of 7, or how many different ways the two dice will land, it will be useful to consider all the possible outcomes. The set of all outcomes of an experiment is called the *sample space* of the experiment. In order to keep track of the elements of the sample space in this experiment, let the first die be green and the second die be red. Since the green die can fall in one of six ways, and the red die can fall in one of six ways, there should be $6 \cdot 6$ or 36 elements in the sample space. The elements of the sample space are as follows:

green	red	green	red	green	red	green	red	green	red	green	red
1	1	2	1	3	1	4	1	5	1	6	1
1	2	2	2	3	2	4	2	5	2	6	2
1	3	2	3	3	3	4	3	5	3	6	3
1	4	2	4	3	4	4	4	5	4	6	4
1	5	2	5	3	5	4	5	5	5	6	5
1	6	2	6	3	6	4	6	5	6	6	6

The circled elements of the sample space are those whose sum is 7.

$$P(7) = \frac{\text{number of successes}}{\text{total number}} = \frac{6}{36} = \frac{1}{6}.$$

The probability, p, of any event is a number such that $0 \le p \le 1$. If $p = 0$, the event cannot happen. If $p = 1$, the event is sure to happen.

Example 4

(A) What is the probability of getting a 7 when one die is thrown?
(B) What is the probability of getting a number less than 12 when one die is thrown?

SOLUTIONS

(A) $P(7) = 0$ since a single die has only numbers 1 through 6 on its faces.

(B) $P(\# < 12) = 1$ since any face number is less than 12.

The *odds* in favor of an event happening are defined to be the probability of the event happening successfully divided by the probability of the event not happening successfully.

Example 5

What are the odds in favor of getting a number greater than 2 when one die is thrown?

$P(\# > 2) = \dfrac{4}{6} = \dfrac{2}{3}$ and $P(\# \ngtr 2) = \dfrac{2}{6} = \dfrac{1}{3}$. Therefore, the odds in favor of a number greater than

$$2 = \frac{P(\# > 2)}{P(\# \ngtr 2)} = \frac{\frac{2}{3}}{\frac{1}{3}} = \frac{2}{1} \quad \text{or} \quad 2:1.$$

INDEPENDENT EVENTS

Independent events are events that have no effect on one another. Two events are defined to be independent if and only if $P(A \cap B) = P(A) \cdot P(B)$, where $A \cap B$ means both events A and B happen. If two events are not independent, they are said to be *dependent*.

Example 1

If two fair coins are flipped, what is the probability of getting two heads?

Since the flip of each coin has no effect on the outcome of any other coin, these are independent events.

$$P(\text{HH}) = P(\text{H}) \cdot P(\text{H}) = \frac{1}{2} \cdot \frac{1}{2} = \frac{1}{4}.$$

Example 2

When two dice are thrown, what is the probability of getting two 5s?

These are independent events because the result of one die does not affect the result of the other.

$$P(\text{two 5s}) = P(5) \cdot P(5) = \frac{1}{6} \cdot \frac{1}{6} = \frac{1}{36}$$

Example 3

Two dice are thrown. Event *A* is "the sum of 7." Event *B* is "at least one die is a 6." Are A and B independent?

From the chart in Example 3, $A = \{(1,6), (6,1), (2,5), (5,2), (3,4), (4,3)\}$ and

$B = \{(1,6), (2,6), (3,6), (4,6), (5,6), (6,6), (6,1), (6,2), (6,3), (6,4), (6,5)\}$.

Therefore, $P(A) = \frac{1}{6}$ and $P(B) = \frac{11}{36}$. $A \cap B = \{(1,6)(6,1)\}$. Therefore, $P(A \cap B) = \frac{2}{36} = \frac{1}{18}$.

$P(A) \cdot P(B) = \frac{11}{216} \neq \frac{1}{18}$. Therefore, $P(A \cap B) \neq P(A) \cdot P(B)$, and so events A and B are dependent.

Example 4

If the probability that John will buy a certain product is $\frac{3}{5}$, that Bill will buy that product is $\frac{2}{3}$, and that Sue will buy that product is $\frac{1}{4}$, what is the probability that at least one of them will buy the product?

Since the purchase by any one of the people does not affect the purchase by anyone else, these events are independent. The best way to approach this problem is to consider the probability that none of them buys the product.

Let A = the event "John does not buy the product."
Let B = the event "Bill does not buy the product."
Let C = the event "Sue does not buy the product."

> **TIP**
>
> To find the probability of "at least one," find 1 – probability of "none."

$$P(A) = 1 - \frac{3}{5} = \frac{2}{5}; \quad P(B) = 1 - \frac{2}{3} = \frac{1}{3}; \quad P(C) = 1 - \frac{1}{4} = \frac{3}{4}$$

The probability that none of them buys the product = $P(A \cap B \cap C) =$

$P(A) \cdot P(B) \cdot P(C) = \frac{2}{5} \cdot \frac{1}{3} \cdot \frac{3}{4} = \frac{1}{10}$. Therefore, the probability that at least one of

them buys the product is $1 - \frac{1}{10} = \frac{9}{10}$.

MUTUALLY EXCLUSIVE EVENTS

In general, the probability of event A happening or event B happening or both happening is equal to the sum of $P(A)$ and $P(B)$ less the probability of both happening. In symbols, $P(A \cup B) = P(A) + P(B) - P(A \cap B)$, where $A \cup B$ means the union of sets A and B. If $P(A \cap B) = 0$, the events are said to be *mutually exclusive*.

Example 1

What is the probability of drawing a spade or a king from a deck of 52 cards?

Let A = the event "drawing a spade."
Let B = the event "drawing a king."

Since there are 13 spades and 4 kings in a deck of cards,

$$P(A) = \frac{13}{52} = \frac{1}{4}; \quad P(B) = \frac{4}{52} = \frac{1}{13}$$

TIP
Generally in probability, "and" means multiply and "or" means add.

$$P(A \cap B) = P(\text{drawing the king of spades}) = \frac{1}{52}$$

$$P(A \cup B) = P(A) + P(B) - P(A \cap B)$$
$$= \frac{13}{52} + \frac{4}{52} - \frac{1}{52} = \frac{16}{52} = \frac{4}{13}.$$

These events are *not* mutually exclusive.

Example 2

In a throw of two dice, what is the probability of getting a sum of 7 or 11?

Let A = the event "throwing a sum of 7."
Let B = the event "throwing a sum of 11."

$P(A \cap B) = 0$, and so these events *are* mutually exclusive.
$P(A \cup B) = P(A) + P(B)$. From the chart in Example 3,

$$P(A) = \frac{6}{36} \quad \text{and} \quad P(B) = \frac{2}{36}.$$

Therefore,

$$P(A \cup B) = \frac{6}{36} + \frac{2}{36} = \frac{8}{36} = \frac{2}{9}.$$

EXERCISES

1. With the throw of two dice, what is the probability that the sum will be a prime number?

 (A) $\dfrac{4}{11}$

 (B) $\dfrac{7}{18}$

 (C) $\dfrac{5}{12}$

 (D) $\dfrac{5}{11}$

 (E) $\dfrac{1}{2}$

2. If a coin is flipped and one die is thrown, what is the probability of getting a head or a 4?

 (A) $\dfrac{1}{12}$

 (B) $\dfrac{1}{3}$

 (C) $\dfrac{5}{12}$

 (D) $\dfrac{7}{12}$

 (E) $\dfrac{2}{3}$

3. Three cards are drawn from an ordinary deck of 52 cards. Each card is replaced in the deck before the next card is drawn. What is the probability that at least one of the cards will be a spade?

 (A) $\dfrac{3}{52}$

 (B) $\dfrac{9}{64}$

 (C) $\dfrac{3}{8}$

 (D) $\dfrac{37}{64}$

 (E) $\dfrac{3}{4}$

4. A coin is tossed three times. Let A = {three heads occur} and B = {at least one head occurs}. What is $P(A \cup B)$?

(A) $\dfrac{1}{8}$

(B) $\dfrac{1}{4}$

(C) $\dfrac{1}{2}$

(D) $\dfrac{3}{4}$

(E) $\dfrac{7}{8}$

5. A class has 12 boys and 4 girls. If three students are selected at random from the class, what is the probability that all will be boys?

(A) $\dfrac{1}{55}$

(B) $\dfrac{1}{4}$

(C) $\dfrac{1}{3}$

(D) $\dfrac{11}{28}$

(E) $\dfrac{11}{15}$

6. A red box contains eight items, of which three are defective, and a blue box contains five items, of which two are defective. An item is drawn at random from each box. What is the probability that both items will be nondefective?

(A) $\dfrac{3}{20}$

(B) $\dfrac{3}{8}$

(C) $\dfrac{5}{13}$

(D) $\dfrac{8}{13}$

(E) $\dfrac{17}{20}$

7. A hotel has five single rooms available, for which six men and three women apply. What is the probability that the rooms will be rented to three men and two women?

(A) $\dfrac{23}{112}$

(B) $\dfrac{97}{251}$

(C) $\dfrac{10}{21}$

(D) $\dfrac{5}{9}$

(E) $\dfrac{5}{8}$

8. Of all the articles in a box, 80% are satisfactory, while 20% are not. The probability of obtaining exactly five good items out of eight randomly selected articles is

(A) 0.003

(B) 0.013

(C) 0.132

(D) 0.147

(E) 0.800

Answers and Explanations

In these solutions the following notation is used:

  i: calculator unnecessary
  a: calculator helpful or necessary
  g: graphing calculator helpful or necessary

Probability

1. i **(C)** There is 1 way to get a 2, and there are 2 ways to get a 3, 4 ways to get a 5, 6 ways to get a 7, 2 ways to get an 11. Out of 36 elements in the sample space, 15 successes are possible.

$$P(\text{prime}) = \frac{15}{36} = \frac{5}{12}.$$

2. i **(D)** The probability of getting neither a head nor a 4 is $\dfrac{1}{2} \cdot \dfrac{5}{6} = \dfrac{5}{12}$. Therefore,

probability of getting either is $1 - \dfrac{5}{12} = \dfrac{7}{12}$.

3. a **(D)** Since the drawn cards are replaced, the draws are independent. The probability

that none of the cards was a spade $= \dfrac{39}{52} \cdot \dfrac{39}{52} \cdot \dfrac{39}{52} = \dfrac{3}{4} \cdot \dfrac{3}{4} \cdot \dfrac{3}{4} = \dfrac{27}{64}$.

Probability that 1 was a spade $= 1 - \dfrac{27}{64} = \dfrac{37}{64}$.

4. i **(E)** The only situation when neither of these sets is satisfied occurs when three tails appear. $P(A \cup B) = \frac{7}{8}$.

5. a **(D)** There are 16 students altogether. The probability that the first person chosen is a boy is $\frac{12}{16}$. Now there are only 15 students left, of which 11 are boys, so the probability that the second student chosen is also a boy is $\frac{11}{15}$. By the same reasoning, the probability that the third is a boy is $\frac{10}{14}$. Therefore, the probability that the first and the second and the third students chosen are all boys is $\frac{12}{16} \times \frac{11}{15} \times \frac{10}{14} = \frac{11}{28}$.

6. i **(B)** Probability of both items being nondefective $= \frac{5}{8} \cdot \frac{3}{5} = \frac{3}{8}$.

7. a **(C)** $\binom{6}{3}$ is the number of ways 3 men can be selected. $\binom{3}{2}$ is the number of ways 2 women can be selected. $\binom{9}{5}$ is the total number of ways people can be selected to fill 5 rooms.

$$P(3 \text{ men, } 2 \text{ women}) = \frac{\binom{6}{3}\binom{3}{2}}{\binom{9}{5}} = \frac{10}{21}.$$

8. a **(D)** Since the problem doesn't say how many articles are in the box, we must assume that it is an unlimited number. The probability of picking 5 satisfactory items (and therefore 3 unsatisfactory ones) is $(0.8)^5(0.2)^3$, and there are $\binom{8}{5}$ ways of doing this.

Therefore, the desired probability is $\binom{8}{5}(0.8)^5(0.2)^3 \approx 0.147$.

PART 3

GRAPHING CALCULATORS

Introduction

• Graphing Calculator Basics	• Typical Questions for Solving

This section will help you use the power of a graphing calculator to improve your score on the Level 2 Subject Test. If you have never used a graphing calculator, there is enough descriptive detail to give you the basics of its operation and guide you through the specific applications. If you are an experienced user, this section serves as a reference for using a graphing calculator to solve the kinds of problems you will encounter.

Although there are several manufacturers and models of graphing calculators on the market, we are most familiar with the TI-83/TI-84 families (including the Plus and Silver Edition models), so we adopted this model for instructional purposes. Users of this book who have calculators made by other manufacturers should have little difficulty finding the keystrokes on their calculators that correspond to those described for the TI-83.*

The chapters that follow provide keystroke detail on the TI-83's general features, as well as the methods that are used to solve specific types of problems. Screen images that result from these keystrokes are also displayed to help users who have calculators in hand to check the accuracy of their work. Readers may nonetheless wish to consult their User's Manual to supplement the instruction given in this book.

Basic features and operations of the calculator are covered in Chapter 7. The logic of the arrangement of the keys is outlined first. The main types of screens that you will use to do the work are described next. A brief overview of the computational logic of the calculator is given next, followed by tips on editing entries into the calculator. Selections of computational mode are described, with recommendations for the Math Level 2 Subject Test. The menu keys and the variety of functions listed within these menus are outlined next. The final section describes "support functions" that facilitate the use of the TI-83.

Chapter 8 provides keystroke detail for using the TI-83 to solve a broad range of problems that might be encountered on the Math Level 2 Subject Test. The chapter is organized around a group of questions that contain descriptive material supported by 20 examples. For convenience, the questions are listed below:

- How do I find a good window for a graph or group of graphs?
- How do I evaluate a function at a specific value?
- How do I find the real zeros of a function?
- How do I find a relative maximum or minimum of a function?
- How do I graph an equation that is not a function?
- How do I use graphs to solve equations?
- How do I use tables to locate the zeros of a polynomial?

---

*Excluded are newer generation graphing calculators with computer algebra systems, such as the TI-89.

- How can I use a graph to solve a polynomial inequality?
- How do I use a graph to check for symmetry?
- How do I use a table to find the limit of a rational function?
- How do I use graphs to solve problems involving absolute value and greatest integer functions?
- How do I generate a sequence?
- How can I find the sum, product, or mean?
- How do I use the factorial, permutation, and combination commands?
- How do I use the calculator to iterate a function?
- How do I graph parametric equations on my calculator?

Finally, Chapter 8 includes five short, simple programs for formulas you may want to use on the Math Level 2 Subject Test:

- The distance between two points
- The midpoint of a segment
- The distance between a point and a line
- The angle between two lines in the plane
- The Quadratic Formula

Screens showing the program commands and screens showing the execution and output of these programs are included.

Basic Operations

CHAPTER 6

- Keyboard Basics
- Review of Main Screens
- Computation and Editing Capabilities

- Menu Keys
- Support Functions

KEYS

The photo on the next page shows the TI-83 calculator keyboard. The top row (right under the screen) contains the keys that govern graphing and tabulation. The four arrow keys move the cursor around the screen. The STAT, MATH, MATRX, PRGM, and VARS keys lead to menus of commands in those categories. The 2nd key and ALPHA key activate the corresponding functions imprinted above the keys. When referring to 2nd or ALPHA prefixes, the prefix itself will not be used after its first mention. For example, after its first mention, 2nd CALC will just be called CALC.

SCREENS

There are three main "screens" that come into play: the Home Screen, where computations are made; the Y = Screen, where formulas are defined; and the Graphing Screen, where graphs are displayed. Sample screens are shown below.

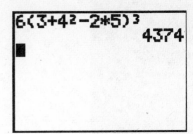

Home Screen

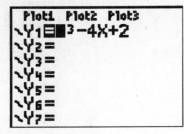

Y = Screen

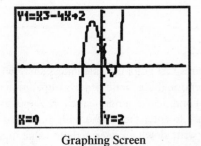

Graphing Screen

You can return to the Home Screen from either of the other two screens by keying 2nd QUIT. You may also use the 2nd TABLE and STAT/EDIT/Edit screens to solve problems. The TABLE screen displays functional values in tabular, rather than graph form. This format may be more convenient, for example, when trying to locate a zero between two integers. The STAT/EDIT/Edit screen, on the other hand, provides a spreadsheet capability. You can do arithmetic on a list or you can combine lists arithmetically. Sample TABLE and STAT/EDIT/Edit screens are shown below.

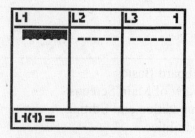

Tabulated Values of $x^2 - 5x + 3$ Blank STAT/EDIT/Edit Screen

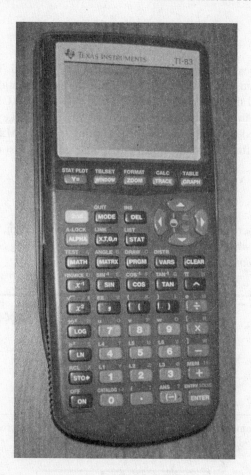

COMPUTATION

There are two important features that graphing calculators share with the newer scientific calculators. Both follow the conventional algebraic order of operations, including parentheses, and both can be used to evaluate complicated numerical expressions. Both also have implied multiplication (i.e., it is unnecessary to enter the × key) with the use of parentheses. With these features it is rarely necessary to write down an intermediate value in a multistep expression.

EDITING

Editing capabilities are another feature of graphing calculators that is shared by newer scientific calculators. On the TI-83, editing can be done using the DEL (delete) and 2nd INS (insert) key. If you make an error while making an entry into the calculator, simply overwrite or delete the error and insert the correct expression.

The 2nd ENTRY key (above ENTER) pastes previous Home Screen entries back into the Home Screen for editing—either to correct errors or to make small changes to the computation. For example, if you evaluate an expression on the Home Screen and need to perform the same operation again, but with one different value, you need not rekey the entire command; simply key ENTRY, and edit the command as desired. This feature is especially useful if the original command involves multiple keystrokes.

The 2nd ANS key can also be used as an edit tool. Suppose, for example, that you want to compute a quantity, say $4 + 8(3 - 17) = -108$ and you realize that you want to divide 45 by this quantity. Instead of rekeying -108, you can just key $45 \div$ ANS as shown below. This is particularly useful if you don't want to round off the ANS value to rekey it. Another convenient feature changes (decimal) answers to fractions. You can access this feature by keying MATH, followed by ENTER twice. This activates the MATH menu and applies the command to change the answer to a fraction ($\triangleright$ Frac).

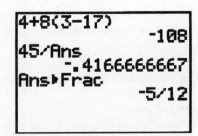

Sample Use of ANS and ANS $\triangleright$ FRAC

GRAPHING COMMANDS

The five keys just below the screen govern the graphing features of the TI-83. Formulas for functions are entered via the Y = key. The WINDOW key is used to set ranges and scales for the coordinate axes. The ZOOM key enables you to zoom in or out of a graph or to select prespecified ranges and scales. Use the TRACE key to locate points on a graph by the simultaneous flashing of the cursor on the graph and giving you its x and y co-ordinates. You can trace a graph using the left and right arrows. If you have more than one graph on a screen, you can move from one to another by using the up and down arrows. The GRAPH key produces a graph or returns to the graphing screen if a graph has already been produced. Examples of specific graphing commands are given in the next chapter.

MODE

The MODE key gives you choices for several of the calculator's important features. The leftmost choice on each line is the default choice. The MODE screen is shown below.

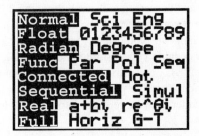

Default Setting on MODE Screen

You alway should use the Normal and Float modes on the Math Level IIC Test. When working a trigonometry problem, you must decide whether to use Radian or Degree measure. You will use Function mode for most problems, but you occasionally will need to change to Polar or Parametric mode for these types of problems.

When graphing, pixels are connected by segments in Connected mode, but they are simply darkened when in Dot mode. Only use Dot mode when you know that there are discontinuities in the graph, as these will be concealed in Connected mode. When graphing two or more functions on the same screen, Sequential mode graphs them in sequence while Simul mode graphs them simultaneously. There is rarely good reason to choose Simul mode over Sequential mode.

Real mode is appropriate for most problems, but it does not display imaginary number results unless imaginary numbers are input. If you get an imaginary number result with real inputs while in Real mode, you will get an error message. On the other hand, $a + bi$ mode displays imaginary number results when they occur, even if the inputs are not imaginary. For example $\sqrt{-36}$ would result in $6i$. You should use $a + bi$ mode on the Math Level 2 Subject Test.

MENU KEYS

As noted above, the menu keys STAT, MATH, MATRX, PRGM, VARS, and their 2nd functions LIST, TEST, ANGLE, DRAW, and DISTR offer a vast array of specialized capabilities in these categories. The submenus and functions you will need to use on the Math Level 2 Subject Test are summarized below. The use of these functions is described in the next section on specific applications.

MATH/MATH:	>Frac—changes decimal values to fractional form
MATH/NUM:	abs(—absolute value
	int(—greatest integer
MATH/PRB:	nPr—number of permutations of r out of n objects
	nCr—number of combinations of r out of n objects
	!—factorial
VARS/Y-VARS:	Function—pastes Y-defined functions from the Y = Screen
	to the Home Screen
LIST/OPS:	seq(—generates a user-defined sequence of numbers
LIST/MATH:	mean(—finds the mean of a list of numbers
LIST/MATH:	sum(—finds the sum of the numbers in a list
LIST/MATH:	prod(—finds the product of the numbers in a list

SUPPORT FUNCTIONS

You can adjust the contrast (make the screen darker or lighter) by keying 2nd up arrow or 2nd down arrow, respectively. When you do this, a digit is briefly displayed in the upper right corner of the screen. With fresh batteries, this digit should be about 2 when the screen is at the right contrast. If the digit is 7 or more when the contrast isn't sufficient, you should consider changing the battery soon. (There is a "change battery" warning when you really must change the battery.)

You can also "jump down" or "jump up" a list of menus, programs, or data by keying ALPHA down arrow or ALPHA up arrow, respectively. Programs are listed in alphabetical order. If, for example, you want to execute the QUAD (Quadratic Formula) program from a long list of programs, simply key ALPHA Q and the screen will scroll through all the programs directly to those beginning with the letter Q. Or, if you need to edit the 23rd number in a list, rather than arrow down and scroll through the first 22 numbers, you can get there faster by keying ALPHA down arrow.

The STO and 2nd RCL keys enable you to store values into the variables ALPHA A through Z, or into lists 2nd L_1 through 2nd L_6, and retrieve them. You should not need to store individual values into the ALPHA locations to solve problems on the Math Level IIC Test, but it may at times be necessary to store sequences you generate into lists. More on this can be found in the specific applications that follow.

Specific Applications

CHAPTER 7

- Basic Features
- Operations
- Types of Screens
- Menu Keys
- Functions

This section, written in question-and-answer format, describes specific TI-83 applications that are on the Math Level 2 Subject Test. The decision whether to use a graphing calculator on a particular test item depends on (a) your experience with the graphing calculator, especially in the course or courses you took where the topic was taught; and (b) the nature of the answer choices for the test item.

As a general rule, when answer choices are exact (such as square roots or expressions involving π), you are expected to solve the problem without a calculator (graphing or scientific). You should assess your ability to do this and use a calculator to obtain a decimal approximation only if you decide that you cannot otherwise solve the problem. You can then use your calculator to evaluate answer choices until a match is achieved. If you are able to formulate a graphing calculator solution immediately, this strategy may be best, even if it means evaluating exact answer choices one at a time until the correct one is found.

If the answer choices for a test item are all in decimal form, you are most likely expected to use a scientific calculator or graphing calculator to arrive at the correct answer choice. Again, if most of your classroom experience is with symbolic (noncalculator) strategies, your best approach is to attempt to solve the problem this way until you reach a calculator-ready answer. Then, use your calculator to evaluate the decimal form of your answer to find the correct answer choice. However, if your classroom experience was based on graphical and tabulation strategies to solve problems, your graphing calculator should give you the correct answer choice more quickly and efficiently than will a symbolic approach. This is the strategy that is addressed primarily in this section.

A few words about graphing calculator syntax are in order before beginning the specific applications. Calculator syntax refers to the structure of calculator commands. These are not always the same as when written algebraically. For example, the mathematical expression $\sin^2 x$ is written as $\sin(x)^2$ in calculator syntax. Also, several of the menu commands, such as absolute value, include a left parenthesis: Abs(, and the user must remember to key the right parenthesis to conclude this command, especially if it is part of a more complex expression. Parentheses that are not necessary in algebra are used in calculator syntax. For example, an algebraic expression such as $\frac{5+6}{3+4}$ does not need parentheses because the division bar serves as a grouping symbol, but when this expression is entered into the calculator, it must be entered as $(5 + 6)/(3 + 4)$.

Graphing calculators are not infallible. Their algorithms for locating points of intersection, zeros, and relative extrema can break down if a function is not well-behaved near the x coordinate of the point in question. For example, the algorithm may not be able to find a zero if it occurs at a point where a function has a vertical or near-vertical tangent line. Fortunately, it is unlikely that such situations would be encountered on the Math Level 2 Test.

How Do I Find a Good Window for a Graph or Group of Graphs?

A good window is defined as one that captures the features of a graph or several graphs that enable you to answer a question. The main features of a graph are its turning points, vertical asymptotes, x and y intercepts, and "end behavior" (the behavior of y as x approaches positive or negative infinity). A good window for the graphs of more than one function should include their point or points of intersection.

Unfortunately, there is no single formula for finding a good window. Practice, to gain experience, is the best way to identify good windows. If an equation(s) contain(s) "small" numerical values, ZStandard (the standard window), which gives a –10 to 10 window in both x and y directions, is a good starting point. Depending on what you see, you can Zoom In or Zoom Out to get a clearer picture of the graph's essential features. In other cases, you must evaluate the function or functions at one or more values to set the window manually. Some examples are given below.

Example 1

Where does $P(x) = x^3 + 18x - 30$ have a zero?

If you only need one zero, ZStandard will do the job (one zero at $x \approx 0.48$), as shown in the window below.

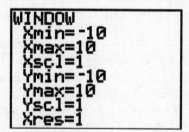 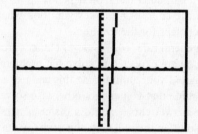

If you need all of the zeros, however, you need to establish whether the graph crosses the x axis at some other point(s). One approach is to take a very "large" window and graph the function in this window, as shown below.

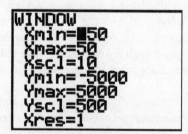 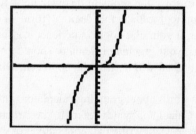

It looks as if there is only one zero, but you should Zoom In several times and adjust the window manually to convince yourself.

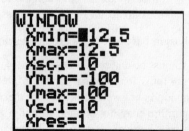 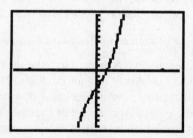

Example 2

Find the maximum value of the function $f(x) = \sqrt{1 + \sin^2 x}$ on the interval $\left[-\dfrac{\pi}{4}, \dfrac{\pi}{4}\right]$.

This problem is simpler than the one in the previous example because the problem specifies the x values to be considered. Key $-\pi/4$, $\pi/4$, and $\pi/8$ for Xmin, Xmax, and Xscl, respectively. Note that the calculator automatically converts these values to decimal form. If you remember that $\sin x \le 1$, so $1 + \sin^2 x \le 2$, you could deduce $0 \le \sqrt{1 + \sin^2 x} \le 2$, giving you Ymin and Ymax values of 0 and 2, respectively. Otherwise, start Ymin and Ymax at -10 and 10, and adjust as necessary. The window and graph for this example are shown below.

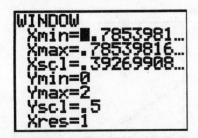 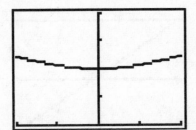

You can see from the graph that the maximum occurs at the endpoints, $-\dfrac{\pi}{4}$ and $\dfrac{\pi}{4}$. Thus, the maximum value of approximately 1.225 can be found by tracing to either of these endpoints.

Example 3

What is the maximum value of $6 \sin x \cos x$?

A good place to start with the window for this function is ZTrig, which automatically sets an $x \varepsilon [-2\pi, 2\pi]$ and $y \varepsilon [-4, 4]$ window.* As can be seen from the graph below, this proves to be sufficient.

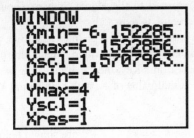 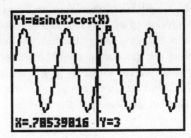

The maximum value of 3 for this function occurs when $x \approx 0.785$.

---

*You frequently will see the phrase "Plot the graph of . . . in an $x \varepsilon [a,b]$ and $y \varepsilon [c,d]$ window" throughout the book. The variables a, b, c, d stand for Xmin, Xmax, Ymin, and Ymax, respectively.

Example 4

Find the area of the region enclosed by the graph of the polar curve

$$r = \frac{1}{\sin\theta + \cos\theta} \text{ and the } x \text{ and } y \text{ axes.}$$

First, set the MODE to Pol[ar]. Unless you know what the polar graph looks like, start again with ZStandard, which sets θ min and θ max at 0 and 2π, respectively, with 48 increments of 0.131 (θ step) for polar graphs. The other parameters are the same as when function graphing with ZStandard. The result of this graph is a line, as indicated in the far left figure below. The area described in the problem is too small to see, so adjust the window settings: Xmin = 0, Xmax = 2, Xscl = 0.5, Ymin = 0, Ymax = 2, Yscl = 0.5 to get the graph shown in the far right window.

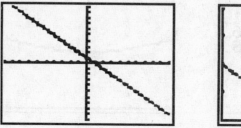 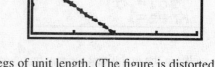

The region is an isosceles right triangle with legs of unit length. (The figure is distorted because the screen isn't square.) The area of this triangle is 0.5.

How Do I Evaluate a Function at a Specific Value?

You have many options to accomplish function evaluation. The best choice depends on the nature and context of the problem.

- If the formula for the function is fairly simple, and you need only one value, simply key in the formula on the Home Screen, with the numerical value in place of x.
- If the formula for the function is complex, key it into one of the Y = lines. Then key VARS/Y-VARS/Function/Y_k and Y_k will be returned to the Home Screen. Then key (x value) followed by ENTER, and the functional value is returned.
- Another option is to key the function formula into one of the Y = lines as above, plot the graph of the function in a suitable window, and key 2nd CALC/value, which returns to the Graphing Screen with X = displayed at the bottom. Key the desired value of X; key ENTER, and the cursor is shown on the graph at that value of X, with both X and Y values displayed at the bottom of the screen.

Screens for these options to find $f(2.6)$ for the function $f(x) = x^3 - 6x + 2$ are shown below.

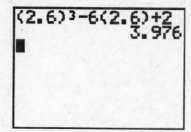 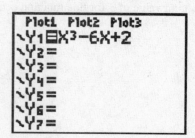

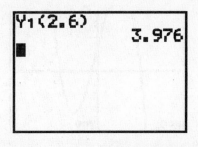

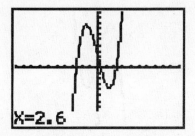

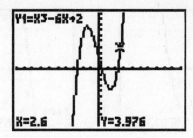

All three options produce the answer 3.976.

Example 5

What is the remainder when $3x^3 + 2x^2 - 5x - 8$ is divided by $x + 2$?

The Remainder Theorem tells you that the remainder in this case is the polynomial evaluated when $x = -2$. Although it requires more keystrokes, the safest method is to enter the polynomial into Y_1, return to the Home Screen, and enter $Y_1(-2)$.

How Do I Find the Real Zeros of a Function?

The real zeros of a function are the x intercepts of its graph. First, plot a graph of the function that includes the x intercept(s) of interest. Key CALC/zero. The calculator asks for a left bound, so move the cursor to some point on the graph to the left of the x intercept and key ENTER. The cursor then asks for a right bound, so move the cursor to some point on the graph to the right of the x intercept and key ENTER. Next, the calculator prompts Guess, so put the cursor as close to the x intercept as you can, and key ENTER again. The coordinates of the x intercept (where the y coordinate is either 0 or a number very close to 0) are then displayed.

Of course, functions can have more than one real zero. As illustrated in the example below, the problem must provide enough information to let you know which zero you need to find.

Example 6

Find the smallest positive real zero of $y = x^3 - 6x + 2$.

Plot the graph of $y = x^3 - 6x + 2$ in the standard window. Although not absolutely necessary, Zoom In to get a better view of the smaller of the two positive zeros. Key CALC/zero and select points to the left and right of this point. Note the two pointers indicating these left and right bounds. Finally, move the cursor as close as you can to the zero and key ENTER. The sequence of screens is shown below.

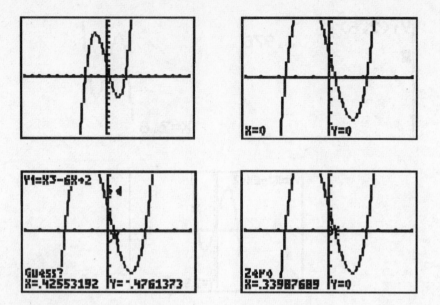

The far right screen shows the smaller zero $x \approx 0.340$.

How Do I Find a Relative Maximum or Minimum of a Function?

Relative minima and maxima of a function are turning points on its graph. The procedure for finding these values is very similar to finding a zero of a function. This time, the relative maximum value must appear in the graph. After keying CALC/maximum, select points to the left and right of the relative maximum. Place the cursor as close to the relative maximum as possible, and key ENTER. The x and y coordinates of the relative maximum will be displayed at the bottom of the screen. Use the same procedure with the MINIMUM key to locate a relative minimum.

Example 7

If $0 \leq x \leq 2\pi$, for what value of x does the function $\sin \dfrac{1}{3}x$ achieve its maximum?

Graph $y = \sin((1/3)x)$ in an $x\varepsilon[0,2\pi]$ and $y\varepsilon[-1,1]$ window. Key CALC/maximum and ENTER left and right bounds of the maximum point. Scroll to your guess for the maximum point and key ENTER. The x and y coordinates of the maximum point will be displayed at the bottom of the screen, as illustrated by the screens below. The function achieves its maximum of 1 when $x \approx 4.712$.

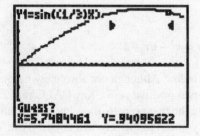
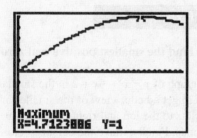

How Do I Graph an Equation That Is Not a Function?

Not all equations are or can be transformed into equations that define functions. For example, the equation of a circle, $x^2 + y^2 = 25$, defines two functions: $y = \sqrt{25 - x^2}$ and $y = -\sqrt{25 - x^2}$, corresponding to the top and bottom halves of the circle. Other conic sections share this characteristic. On the Math Level 2 Test, the main reason you need to be able to graph an equation that is not a function is to answer questions about symmetry. Therefore, you will have to do some algebra to solve for y, typically as $y = \pm f(x)$. Enter the equation for $f(x)$ into y_1; then enter $-y_1$ into y_2 and graph both functions.

Example 8

Assess the symmetry of $x^2 - y^2 = 1$. Solving for y, $y^2 = x^2 - 1$, so $y = \pm\sqrt{x^2 - 1}$.

First plot the graphs of $y = \sqrt{x^2 - 1}$ and $y = -y_1$ in an $x \varepsilon [-2,2]$ and $y \varepsilon [-2,2]$ window. Since you are using the visual image to draw conclusions about the equation, you need to key ZOOM/Zsquare to avoid distortion. The screens below show that the equation is symmetrical about both axes and the origin.

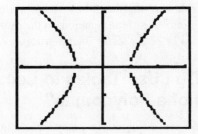

How Do I Use Graphs to Solve Equations?

First, you need to enter the two sides of the equation as functions, say one into Y_1 and the other into Y_2. Solutions to the equation are points where these two graphs intersect. As was the case with the zeros of a function, a problem must specify which of several possible solutions it's asking you to find. Once you've determined which point of intersection gives the solution, you must first find a window that shows the point of intersection clearly. Then key CALC/intersect, which returns you to the Graphing Screen. The calculator asks you to designate the "first curve." Key ENTER and the cursor jumps to another curve and asks for the "second curve." If there are only two graphs on the screen, key ENTER again, which specifies the other graph as the second curve. (You rarely will be working with more than two graphs, but if there should be more than two on the screen, you must designate which two you are working with.)

The calculator then asks for your "guess," and you move the cursor as close as you can to the point of intersection and key ENTER again. After a moment, the cursor moves, if necessary, to the point of intersection and the x and y coordinates of the point of intersection are displayed at the bottom of the screen.

Example 9

Solve $\sin 2x = \sin x$ in the interval $(0,\pi)$.

With your calculator in radian mode, plot the graphs of $y = \sin(2x)$ and $y = \sin(x)$ in an $x\varepsilon[0,\pi]$ and $y\varepsilon[-2,2]$ window. The screen below results from keying CALC/intersect, ENTER, ENTER, and moving the cursor close to the point of intersection. The second screen results when ENTER is keyed once again, with the solution $x \approx 1.047$ displayed at the bottom. Note that the two selected curves have + signs on them.

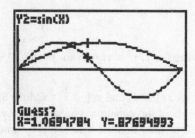
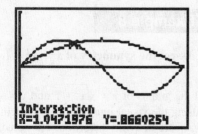

An alternative approach to using graphs to solve an equation is to transform the equation first so that one side is zero. Then follow the procedures for finding a zero of the function that results on the nonzero side of the equation. In the example given above, you would find the zeros of the function $\sin 2x - \sin x$ on the interval $(0,\pi)$.

How Do I Use Tables to Locate the Zeros of a Polynomial?

The Math Level 2 Subject Test may include questions about the location of the zeros of a polynomial. The TABLE feature of the TI-83 provides a way to locate a zero of a polynomial between two integers or, for that matter, between any two values.

Example 10

Between which two consecutive integers is there a zero of $P(x) = 28x^3 - 11x^2 + 15x - 12$?

Enter the formula for $P(x)$ into Y_1, and key 2nd TBLSET. As shown in the screen below, set TblStart to -3 and set ΔTbl to 1. Set both Indepnt and Depend to Auto. This sets the calculator to build a table of values, starting at $x = -3$ and increasing x by 1. There is no particular reason for starting at -3, as once you key 2nd TABLE, you may move up or down by 1 in either the positive or negative direction. The TBLSET and corresponding TABLE screens are shown below.

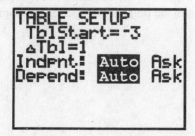
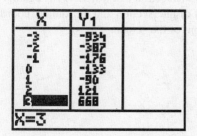

The screen shows that Y_1 ($P(x)$) changes sign between 1 and 2, so you can conclude that the polynomial has a zero between those two integers.

How Can I Use a Graph to Solve a Polynomial Inequality?

Example 11

Solve the inequality $6x^2 - 11x > 7$.

In this problem you have the functions $Y_1 = 6x^2 - 11x$ and $Y_2 = 7$. You need to determine the values of x for which Y_1 lies above Y_2. It is customary, however, to set one side of the inequality equal to zero, $6x^2 - 11x - 7 > 0$, and find the values of x for which this new function lies above the x-axis. Plot the graph of $y = 6x^2 - 11x - 7$ in an $x\varepsilon[-3,5]$ and $y\varepsilon[-15,10]$ window. The graph lies above the x axis to the right of the larger zero and the left of the smaller zero. Using CALC/zero, these values are determined to be

$x = \dfrac{7}{3}$ and $x = -\dfrac{1}{2}$, respectively.

Therefore, the solution consists of all numbers greater than $\dfrac{7}{3}$ or less than $-\dfrac{1}{2}$. The screens for these two zeros are shown below.

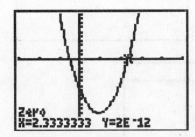 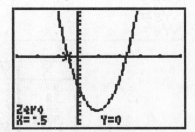

How Do I Use a Graph to Check for Symmetry?

The Math Level 2 Subject Test may ask about three types of symmetry: symmetry with respect to the x- or y-axis, and symmetry with respect to the origin. Symmetry is related to the concepts of even and odd functions. Even functions are symmetric with respect to the y-axis, while odd functions are symmetric with respect to the origin. Whether a function has a certain type of symmetry is evident upon inspection of its graph.

Example 12

What are the symmetries of the function $f(x) = 2x^4 - 3x^2 + 2$?

Plot the graph of this function in the standard window and observe that it is symmetric with respect to the y-axis. The window showing this graph appears below.

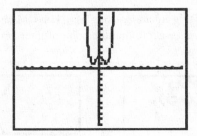

How Do I Use a Table to Find the Limit of a Rational Function?

Example 13

Find the limit of $\dfrac{3x^2 - 3}{x - 1}$ as x approaches 1.

Key $(3x^2 - 3)/(x - 1)$ into Y_1, and key 2nd TBLSET. Select Indpnt: Ask and Depend: Auto, and key TABLE. To let x approach 1, enter values progressively closer to 1, from below and above, as shown in the table screen on the right. It should be pretty clear from the table that the desired limit is 6.

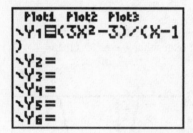
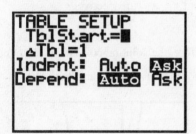

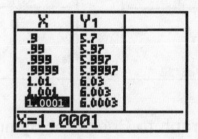

How Do I Use Graphs to Solve Problems Involving Absolute Value and Greatest Integer Functions?

Equations involving absolute value can be very cumbersome to solve algebraically because they often need to be broken down into cases. A graphical approach makes finding solutions much easier.

Example 14

Solve $|2x - 5| = 3x + 4$.

Plot the graph of each side of this equation, using MATH/NUM/abs (on the left side), in the standard window. Use CALC/intersect to find the point of intersection at $x = 0.2$, $y = 4.6$. These screens are shown below.

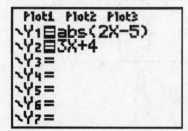
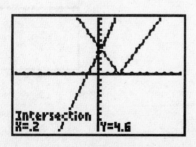

Some of the Math Level 2 Subject Test questions ask you to determine the nth term of a sequence or to find the sum or product of the terms of a sequence. There are formulas for arithmetic and geometric sequences to make these determinations, but you also can use your calculator to generate the sequence and find sums and products.

How Do I Generate a Sequence?

To generate a sequence, use 2nd LIST/OPS/seq, which returns seq(to the Home Screen. The left parenthesis indicates that you must enter some inputs. Four arguments are required: the formula for each term of the sequence; the variable that governs the term; the value of the variable that generates the first term; and the value of the variable that generates the last term. Close with a right parenthesis after these four arguments are entered, and key ENTER. The sequence is listed horizontally in braces.

Since you probably will do something with a sequence (such as finding the sum or mean of the terms), it is best to store the sequence in a list. This can be done either before or after keying ENTER, by keying STO 2nd L_1. You can inspect the sequence in L_1 by keying STAT/EDIT/Edit. (If you don't see L_1, key STAT/EDIT/SetUpEditor, which returns SetUpEditor to the Home Screen. Then key ENTER, followed by STAT/EDIT/Edit again. This time you'll see L_1 on the Home Screen.)

Example 15

Generate the first 12 terms of the geometric sequence 2, 6, 18, 54, ...

The first term of this sequence is 2, and subsequent terms are the product of 2 and increasing powers of 3. We can think $2 = 2 \times 3^0$; $6 = 2 \times 3^1$; $18 = 2 \times 3^2$, and so on. To generate the desired sequence, we need the powers of 3 to go from 0 to 11 (thus giving 12 terms). The command Seq($2 * 3\char94 X$, X,0,11) will generate the desired sequence. (The command Seq($2 * 3\char94 (X - 1)$, X,1,12) would generate the identical sequence.)

If the problem was to find the seventh term, you would store this sequence in L_1 and key 2nd L_1(7). This sequence is displayed in the screen below.

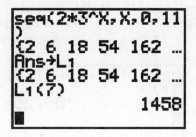

How Can I Find the Sum, Product, or Mean?

A question may ask you to find the sum of the terms of an arithmetic or geometric sequence. Or, a question may ask you to demonstrate your knowledge of sigma notation by finding a sum expressed in that notation. You may be asked to find the product of the terms of a sequence. Finally, you may be asked to find the mean or median of a data set that is too large to easily work with manually. These questions and others can be handled by using menu commands in 2nd LIST/MATH, shown on the screen below.

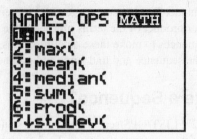

Example 16

Find the sum of the first 20 terms of the arithmetic series 12, 4, –4, ...

Each number in the sequence is 8 less than its predecessor, so create the sequence by keying LIST/OPS/seq($12 - 8x,x,0,19$), or its equivalent, seq($12 - 8(x - 1),1,20$). As noted previously, when you key ENTER, the sequence is displayed horizontally in braces across the Home Screen. Key LIST/MATH/sum(Ans), followed by ENTER, to get the answer –1200. This result can be accomplished in a single step by keying LIST/MATH/sum(seq($12 - 8x,x,0,19$)). These results are displayed on the following screen.

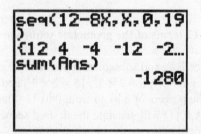

In the example just given, the argument for sum was Ans, in this case a list of numbers in braces { }—2^{nd}( ). Any list presented in this fashion or named (such as L_1, L_2, and so on) can be used as the argument for any of the LIST/MATH commands. For example, if you wanted to find the mean of the 4 numbers 68, 79, 91, 83, simply key LIST/MATH/mean({68,79,91,83}), then ENTER for the result 80.25. This is shown in the following screen.

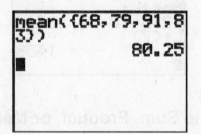

How Do I Use the Factorial, Permutation, and Combination Commands?

These commands are used to solve "combinatorics" problems—the number of ways that a set of objects can be selected and/or arranged. Factorial (!) is used to count the number of permutations (distinct arrangements) of objects. If there are n objects, then there are n! distinct ways in which they can be arranged. Work from the Home Screen to do this calculation. For example, to calculate 6!, key 6, followed by MATH/PRB/!, which returns ! following the 6 on the Home Screen. Key ENTER to get the result 720 (6! = 6*5*4*3*2*1).

If a portion r of n objects is selected, the number of distinct ways these r objects can be arranged is nPr (the number of permutations of r objects of n). For example, to find the number of ways of arranging 6 objects of 10, key 10 in the Home Screen, followed by MATH/PRB/nPr, followed by 6. Then key ENTER to get the result 151200. If the order of the r objects is unimportant—you are simply interested in the number of ways of selecting 6 of 10 objects, regardless of their order— use nCr instead of nPr, to get the result 210. These are shown on the screen below.

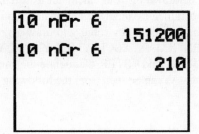

Example 17

How many ways can a president, vice president, and secretary be selected in a club of 20 people?

Since 3 people are being selected from 20, $n = 20$ and $r = 3$. It matters which of the 3 is elected to which office, so use 20 nPr 3 = 6840.

Example 18

How many ways can a committee of 3 people be selected in a club of 20 people?

Since it doesn't matter who is selected to the committee first, second, or third, use 20 nCr 3 = 1140.

How Do I Use the Calculator to Iterate a Function?

You may be asked to iterate a function in an exam problem. In this type of problem you are given a starting value and a rule to move from each value to the next. The problem asks you to find a particular value in such a sequence. Both the TI-83 and the newer scientific calculators are well suited to this type of problem, as illustrated in the following example.

Example 19

Suppose $a_0 = 3$ and $a_{n+1} = 2a_n - 1$. Find a_4.

First, key 3 ENTER. Then key 2 * ANS − 1 ENTER, to get a_1; ENTER again, to get a_2; and ENTER twice more to get the answer a_4. The screen that results from this sequence is shown below.

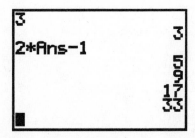

How Do I Graph Parametric Equations on My Calculator?

Recall that parametric equations are a way of representing points on a plane or in space where the coordinates are represented as functions of a parameter. Points in the plane are typically represented as $x(t)$, $y(t)$, where the parameter t represents time, and x and y are coordinates of a point at time t. When it is possible to eliminate the variable t, y can be expressed as a function of x and the graph of the function can be drawn on your calculator as usual.

If, on the other hand, the parameter cannot be eliminated, you must put your calculator in parametric mode and enter the equations for x and y to draw a graph. The window settings in parametric mode include XMIN, XMAX, XSCL, YMIN, YMAX, YSCL as in function mode, as well as the settings TMIN, TMAX, and TSTEP, that define the smallest, largest, and incremental step values for the parameter. As can be seen from the following example, the parameter need not be denoted by t.

Example 20

Sketch the graph of the parametric equations $\begin{cases} x = 2(\theta - \sin\theta) \\ y = 2(1 - \cos\theta) \end{cases}$.

With your calculator in parametric mode, enter the equations for x and y. Since the equations involve trigonometric functions, you need to choose degree or radian mode. If you choose radian mode, a good starting point for TMIN is 0 and for TMAX is 2π. TSTEP determines how smooth the graph will be: the smaller TSTEP, the smoother the graph. A reasonable starting point for TSTEP in this problem is $\dfrac{\pi}{24}$, which will provide 49 data points on the graph. The window settings and resulting graph are shown below.

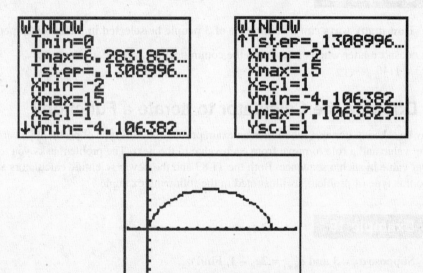

Some experimentation may be necessary to produce a good graphic representation of parametric equations.

Programs

CHAPTER 8

> • Writing the Programs
>
> • Executing Instructions

This chapter describes five programs that you can write for your TI-83 and use as necessary on the Math Level 2 Subject Test. These programs make it easy for you to use common formulas as part of doing larger problems:

- distance between two points (D2P)
- midpoint of a segment (MIDPT)
- distance between a point and a line (DPL)
- angle between two lines in the plane (ANGLE)
- Quadratic Formula (QUADFORM)

First, you will have to write the programs, which are short and easy to write. Once written, the programs can be executed. The two sections that follow explain how to write and execute the five programs listed above.

WRITING THE PROGRAMS

Begin writing each program by keying PRGM/NEW/ENTER. The cursor appears in ALPHA mode following Name =, and you should type the name of the program (using either the name given above or your own choice of names). Key ENTER after typing the name and a colon (:) will appear for you to enter your first program command.

Key PRGM to display submenus CTL, I/O, and EXEC. These stand for control commands, input/output commands, and execution commands (the latter, to execute one program within another). Key I/O to find the commands "Prompt," and "Display," and key CTL to find the command "Stop." When you key these commands, they are pasted into the program, and you key the remaining characters from the calculator's keyboard, using 2nd and ALPHA as needed.

When a program is executed, the characters that are in quotes (" ") are displayed as text. This makes it easier to follow the correct steps when executing the program. The quote sign (") is obtained by keying ALPHA +.

Once written, programs can only be accessed by keying PRGM/EDIT. The five programs are shown on the screens below. Because the programs are relatively short, there is some overlap between screens.

Program D2P

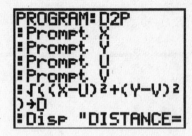

Program MIDPT

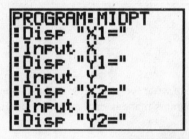

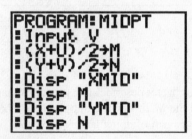

Program DPL

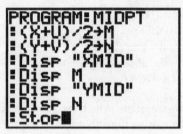

Program ANGLE

Program QUADFORM

EXECUTING THE PROGRAMS

To execute a program you must key PRGM/EXEC and either scroll down to the program you want or enter the number of the program you want. The steps you should follow to run each program are described below, and sample values are shown in the screen output for each one.

D2P

Highlight D2P and key ENTER or key the program number. PrgmD2P will be pasted to the Home Screen. Key ENTER again to begin executing the program. When you see the prompt X? key the *x* coordinate of the point. Prompt Y? for the *y* coordinate of the point is displayed, followed by the prompts U? and V?, for the *x* and *y* coordinates of the second point. After you key the final coordinate and ENTER, the distance is displayed on the screen. This final screen is shown below.

MIDPT

Highlight MIDPT and key ENTER or key the program number. Program MIDPT is pasted to the Home Screen. Key Enter again to begin executing the program. Prompts for X1?, X2?, Y1?, and Y2?, representing the *x* and *y* coordinates of the endpoints of the segment are displayed in sequence. After the final prompt is entered, the midpoint is displayed. Screens to find the midpoint of the point displayed in the previous example are shown below.

 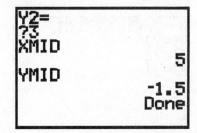

DPL

Highlight DPL and key ENTER or key the program number. Program DPL is pasted to the Home Screen. Key Enter again to begin executing the program. Prompts for X? and Y?, the coordinates of the point, are displayed. The next three prompts are A?, B?, and C?, coefficients of the equation $Ax + By + C = 0$. After the final prompt is entered, the distance is displayed. The screens from the program finding the distance between the point $(3,-5)$ and the line $x - 6y + 9 = 0$ are shown below.

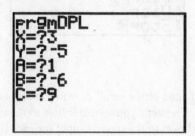

ANGLE

First, be sure to set MODE to "Degree." Highlight ANGLE and key ENTER or key the program number. Program ANGLE is pasted to the Home Screen. This program finds the acute angle formed by two lines in the plane with given slopes. The first prompt M? asks for the slope of one of the lines. Once this value is entered, another prompt N? asks for the slope of the other line. Key ENTER, and the acute angle is given in degrees. A screen showing this sequence appears below.

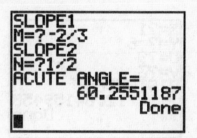

QUADFORM

Highlight QUADFORM and key ENTER or key the program number. Program QUADFORM is pasted to the Home Screen. Prompts for A?, B?, and C?, the coefficients of the equation $Ax^2 Bx + C = 0$, are displayed. After each is entered, the value of D, the discriminant, and the two solutions are displayed. If the $a + bi$ is selected, imaginary solutions are displayed when they occur. Otherwise, the error message "NONREAL ANS" is displayed. The screens for solving $x^2 + x + 1 = 0$ are shown below.

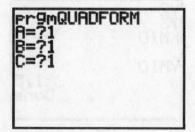

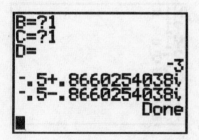

PART 4

MODEL TESTS

Answer Sheet
MODEL TEST 1

1 Ⓐ Ⓑ Ⓒ Ⓓ	14 Ⓐ Ⓑ Ⓒ Ⓓ	27 Ⓐ Ⓑ Ⓒ Ⓓ	40 Ⓐ Ⓑ Ⓒ Ⓓ
2 Ⓐ Ⓑ Ⓒ Ⓓ	15 Ⓐ Ⓑ Ⓒ Ⓓ	28 Ⓐ Ⓑ Ⓒ Ⓓ	41 Ⓐ Ⓑ Ⓒ Ⓓ
3 Ⓐ Ⓑ Ⓒ Ⓓ	16 Ⓐ Ⓑ Ⓒ Ⓓ	29 Ⓐ Ⓑ Ⓒ Ⓓ	42 Ⓐ Ⓑ Ⓒ Ⓓ
4 Ⓐ Ⓑ Ⓒ Ⓓ	17 Ⓐ Ⓑ Ⓒ Ⓓ	30 Ⓐ Ⓑ Ⓒ Ⓓ	43 Ⓐ Ⓑ Ⓒ Ⓓ
5 Ⓐ Ⓑ Ⓒ Ⓓ	18 Ⓐ Ⓑ Ⓒ Ⓓ	31 Ⓐ Ⓑ Ⓒ Ⓓ	44 Ⓐ Ⓑ Ⓒ Ⓓ
6 Ⓐ Ⓑ Ⓒ Ⓓ	19 Ⓐ Ⓑ Ⓒ Ⓓ	32 Ⓐ Ⓑ Ⓒ Ⓓ	45 Ⓐ Ⓑ Ⓒ Ⓓ
7 Ⓐ Ⓑ Ⓒ Ⓓ	20 Ⓐ Ⓑ Ⓒ Ⓓ	33 Ⓐ Ⓑ Ⓒ Ⓓ	46 Ⓐ Ⓑ Ⓒ Ⓓ
8 Ⓐ Ⓑ Ⓒ Ⓓ	21 Ⓐ Ⓑ Ⓒ Ⓓ	34 Ⓐ Ⓑ Ⓒ Ⓓ	47 Ⓐ Ⓑ Ⓒ Ⓓ
9 Ⓐ Ⓑ Ⓒ Ⓓ	22 Ⓐ Ⓑ Ⓒ Ⓓ	35 Ⓐ Ⓑ Ⓒ Ⓓ	48 Ⓐ Ⓑ Ⓒ Ⓓ
10 Ⓐ Ⓑ Ⓒ Ⓓ	23 Ⓐ Ⓑ Ⓒ Ⓓ	36 Ⓐ Ⓑ Ⓒ Ⓓ	49 Ⓐ Ⓑ Ⓒ Ⓓ
11 Ⓐ Ⓑ Ⓒ Ⓓ	24 Ⓐ Ⓑ Ⓒ Ⓓ	37 Ⓐ Ⓑ Ⓒ Ⓓ	50 Ⓐ Ⓑ Ⓒ Ⓓ
12 Ⓐ Ⓑ Ⓒ Ⓓ	25 Ⓐ Ⓑ Ⓒ Ⓓ	38 Ⓐ Ⓑ Ⓒ Ⓓ	
13 Ⓐ Ⓑ Ⓒ Ⓓ	26 Ⓐ Ⓑ Ⓒ Ⓓ	39 Ⓐ Ⓑ Ⓒ Ⓓ	

Model Test 1

T ear out the preceding answer sheet. Decide which is the best choice by rounding your answer when appropriate. Blacken the corresponding space on the answer sheet. When finished, check your answers with those at the end of the test. For questions that you got wrong, note the sections containing the material that you must review. Also if you do not fully understand how you arrived at some of the correct answers, you should review the appropriate sections. Finally, fill out the self-evaluation chart on page 225 in order to pinpoint the topics that give you the most difficulty.

50 questions:1 hour

Directions: Decide which answer choice is best. If the exact numerical value is not one of the answer choices, select the closest approximation. Fill in the oval on the answer sheet that corresponds to your choice.

Notes:
(1) You will need to use a scientific or graphing calculator to answer some of the questions.
(2) You will have to decide whether to put your calculator in degree or radian mode for some problems.
(3) All figures that accompany problems are plane figures unless otherwise stated. Figures are drawn as accurately as possible to provide useful information for solving the problem, except when it is stated in a particular problem that the figure is not drawn to scale.
(4) Unless otherwise indicated, the domain of a function is the set of all real numbers for which the functional value is also a real number.

Reference Information. The following formulas are provided for your information.

Volume of a right circular cone with radius r and height h: $V = \dfrac{1}{3}\pi r^2 h$

Lateral area of a right circular cone if the base has circumference c and slant height is l:

$S = \dfrac{1}{2}cl$

Volume of a sphere of radius r: $V = \dfrac{4}{3}\pi r^3$

Surface area of a sphere of radius r: $S = 4\pi r^2$

Volume of a pyramid of base area B and height h: $V = \dfrac{1}{3}Bh$

1. The slope of a line perpendicular to the line whose equation is $\dfrac{x}{3} - \dfrac{y}{4} = 1$ is

    (A) -3

    (B) $-\dfrac{4}{3}$

    (C) $-\dfrac{3}{4}$

    (D) $\dfrac{1}{4}$

    (E) $\dfrac{4}{3}$

[handwritten: $4x - 3y = 12$]
[handwritten: $4x - 12 = 3y$]
[handwritten: $\frac{4}{3}x - 4 = y$]

USE THIS SPACE FOR SCRATCH WORK

2. What is the range of the data set 8, 12, 12, 15, 18?

    (A) 10
    (B) 12
    (C) 13
    (D) 15
    (E) 18

3. If $f(x) = \dfrac{x-7}{x^2 - 49}$, for what value(s) of x does the graph of $y = f(x)$ have a vertical asymptote?

    (A) -7
    (B) 0
    (C) $-7, 0, 7$
    (D) $-7, 7$
    (E) 7

4. If $f(x) = \sqrt{2x+3}$ and $g(x) = x^2 + 1$, then $f(g(2)) =$

    (A) 2.24
    (B) 3.00
    (C) 3.61
    (D) 6.00
    (E) 6.16

[handwritten: $g(2) = 2^2 + 1 = 5$]
[handwritten: $f(5) = \sqrt{2 \cdot 5 + 3} = \sqrt{13}$]

5. $\left(-\dfrac{1}{16}\right)^{2/3} =$

    (A) -0.25
    (B) -0.16
    (C) 0.16
    (D) 6.35
    (E) The value is not a real number.

6. The circumference of circle $x^2 + y^2 - 10y - 36 = 0$ is

(A) 38
(B) 49
(C) 54
(D) 125
(E) 192

Scratch work:
$x^2 + y^2 - 10y + 25 = 0 + 36 + 25$
$x^2 + (y-5)^2 = 61$

7. What is the value of $\displaystyle\sum_{j=3}^{5} \ln j$?

(A) 1.6
(B) 1.9
(C) 4.1
(D) 4.8
(E) 7.8

8. If $f(x) = 2$ for all real numbers x, then $f(x + 2) =$

(A) 0
(B) 2
(C) 4
(D) x
(E) The value cannot be determined.

9. The volume of the region between two concentric spheres of radii 2 and 5 is

(A) 28
(B) 66
(C) 113
(D) 368
(E) 490

Scratch work: $V = \frac{4}{3}\pi r^3$

10. If a, b, and c are real numbers and if

$$a^5 b^3 c^8 = \frac{9a^3 c^8}{b^{-3}},$$ then a could equal

(A) $\dfrac{1}{9}$

(B) $\dfrac{1}{3}$

(C) 9
(D) 3
(E) $9b^6$

Scratch work:
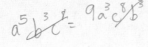
$a^5 = 9a^3$
$a^2 = 9$

Model Test 1

11. In right triangle *ABC*, *AB* = 10, *BC* = 8, *AC* = 6. The sine of ∠*A* is

    (A) $\frac{3}{5}$

    (B) $\frac{3}{4}$

    (C) $\frac{4}{5}$

    (D) $\frac{5}{4}$

    (E) $\frac{4}{3}$

12. If $16^x = 4$ and $5^{x+y} = 625$, then *y* =

    (A) 1

    (B) 2

    (C) $\frac{7}{2}$

    (D) 5

    (E) $\frac{25}{2}$

13. If the parameter is eliminated from the equations $x = t^2 + 1$ and $y = 2t$, then the relation between *x* and *y* is

    (A) $y = x - 1$
    (B) $y = 1 - x$
    (C) $y^2 = x - 1$
    (D) $y^2 = (x - 1)^2$
    (E) $y^2 = 4x - 4$

14. Let *f(x)* be a polynomial function: $f(x) = x^5 + \cdots$. If $f(1) = 0$ and $f(2) = 0$, then *f(x)* is divisible by

    (A) $x - 3$
    (B) $x^2 - 2$
    (C) $x^2 + 2$
    (D) $x^2 - 3x + 2$
    (E) $x^2 + 3x + 2$

15. $\sin\left(\tan^{-1}\frac{1}{3}\right)$ equals

    (A) 0.32
    (B) 0.33
    (C) 0.35
    (D) 0.50
    (E) 0.95

USE THIS SPACE FOR SCRATCH WORK

16. If $z > 0$, $a = z\cos\theta$, and $b = z\sin\theta$, then $\sqrt{a^2 + b^2} =$

(A) 1
(B) z
(C) $2z$
(D) $z\cos\theta\sin\theta$
(E) $z(\cos\theta + \sin\theta)$

$\sqrt{(z\cos\theta)^2 + (z\sin\theta)^2}$

$\sqrt{z^2\cos^2\theta + z^2\sin^2\theta}$
$\sqrt{z^2(\cos^2\theta + \sin^2\theta)}$

17. If the vertices of a triangle are $(u,0)$, $(v,8)$, and $(0,0)$, then the area of the triangle is

(A) $4|u|$
(B) $2|v|$
(C) $|uv|$
(D) $2|uv|$
(E) $\dfrac{1}{2}|uv|$

$\frac{1}{2}u\cdot 8$

18. If $f(x) = \begin{cases} \dfrac{5}{x-2}, & \text{when } x \neq 2 \\ k, & \text{when } x = 2 \end{cases}$, what must the

value of k be in order for $f(x)$ to be a continuous function?

(A) -2
(B) 0
(C) 2
(D) 5
(E) No value of k will make $f(x)$ a continuous function.

19. What is the probability that a prime number is less than 7, given that it is less than 13?

(A) $\dfrac{1}{3}$

(B) $\dfrac{2}{5}$

(C) $\dfrac{1}{2}$

(D) $\dfrac{3}{5}$

(E) $\dfrac{3}{4}$

1 2 3 5 7 11 13

$\dfrac{4}{7}$

20. The ellipse $4x^2 + 8y^2 = 64$ and the circle $x^2 + y^2 = 9$ intersect at points where the y-coordinate is

(A) $\pm \sqrt{2}$

(B) $\pm \sqrt{5}$

(C) $\pm \sqrt{6}$

(D) $\pm \sqrt{7}$

(E) ± 10.00

USE THIS SPACE FOR SCRATCH WORK

$$4(x^2 + 2y^2) = 64$$
$$x^2 + 2y^2 = 16$$
$$- \quad x^2 + y^2 = 9$$
$$y^2 = 7$$

21. Each term of a sequence, after the first, is inversely proportional to the term preceding it. If the first two terms are 2 and 6, what is the twelfth term?

(A) 2

(B) 6

(C) 46

(D) $2 \cdot 3^{11}$

(E) The twelfth term cannot be determined.

$a_n = 2 \cdot \frac{1}{3}^{n-1}$

22. A company offers you the use of its computer for a fee. Plan A costs \$6 to join and then \$9 per hour to use the computer. Plan B costs \$25 to join and then \$2.25 per hour to use the computer. After how many minutes of use would the cost of plan A be the same as the cost of plan B?

(A) 18,052

(B) 173

(C) 169

(D) 165

(E) 157

$A = 6 + 9h$

$B = 25 + 2.25h$

23. If the probability that the Giants will win the NFC championship is p and if the probability that the Raiders will win the AFC championship is q, what is the probability that only one of these teams will win its respective championship?

(A) pq

(B) $p + q - 2pq$

(C) $|p - q|$

(D) $1 - pq$

(E) $2pq - p - q$

$p \cdot (1 - q) \qquad q \cdot (1 - p)$

$p - pq \quad + \quad q - qp$

24. If a geometric sequence begins with the terms $\dfrac{1}{3}, 1, \ldots$, what is the sum of the first 10 terms?

(A) $9841\dfrac{1}{3}$

(B) 6561

(C) $3280\dfrac{1}{3}$

(D) $33\dfrac{1}{3}$

(E) 6

$a_n = \frac{1}{3} \cdot 3^{n-1}$

$S_{10} = \dfrac{\frac{1}{3}(1 - 3^{10})}{1 - 3}$

25. The value of $\dfrac{453!}{450!\,3!}$ is

(A) greater than 10^{100}
(B) between 10^{10} and 10^{100}
(C) between 10^{5} and 10^{10}
(D) between 10 and 10^{5}
(E) less than 10

26. If A is the angle formed by the line $2y = 3x + 7$ and the x-axis, then $\angle A$ equals

(A) $-45°$
(B) $0°$
(C) $56°$
(D) $72°$
(E) $215°$

27. What is the smallest positive x-intercept of the graph

of $y = 3\sin 2\left(x + \dfrac{2\pi}{3}\right)$?

(A) 0
(B) 0.52
(C) 1.05
(D) 1.31
(E) 2.09

28. If $(x - 4)^2 + 4(y - 3)^2 = 16$ is graphed, the sum of the distances from any fixed point on the curve to the two foci is

(A) 4
(B) 8
(C) 12
(D) 16
(E) 32

29. In the equation $x^2 + kx + 54 = 0$, one root is twice the other root. The value(s) of k is (are)

(A) -5.2
(B) 15.6
(C) 22.0
(D) ± 5.2
(E) ± 15.6

30. The remainder obtained when $3x^4 + 7x^3 + 8x^2 - 2x - 3$ is divided by $x + 1$ is

(A) -3
(B) 0
(C) 3
(D) 5
(E) 13

31. If $f(x) = e^x$ and $g(x) = f(x) + f^{-1}(x)$, what does $g(2)$ equal?

    handwritten: $g(x) = e^x + \ln(x)$

    (A) 5.1
    (B) 7.4
    (C) 7.5
    (D) 8.1
    (E) 8.3

32. If $x_0 = 3$ and $x_{n+1} = \sqrt{4 + x_n}$, then $x_3 =$

    handwritten: $X_1 = 2.6$
    X_2

    (A) 2.65
    (B) 2.58
    (C) 2.56
    (D) 2.55
    (E) 2.54

33. For what values of k does the graph of
    $$\frac{(x - 2k)^2}{1} - \frac{(y - 3k)^2}{3} = 1 \text{ pass through the origin?}$$

    (A) only 0
    (B) only 1
    (C) ±1
    (D) ±√5
    (E) no value

    handwritten:
    $(-2k)^2 - \frac{(-3k)^2}{3} = 1$
    $4K^2 - \frac{9k^2}{3} = 1$
    $4K^2 - 3k^2 = 1$
    $K^2 = 1$

34. If $\dfrac{1 - \cos\theta}{\sin\theta} = \dfrac{\sqrt{3}}{3}$, then $\theta =$

    (A) 15°
    (B) 30°
    (C) 45°
    (D) 60°
    (E) 75°

35. If $x^2 + 3x + 2 < 0$ and $f(x) = x^2 - 3x + 2$, then

    (A) $0 < f(x) < 6$

    (B) $f(x) \geq \dfrac{3}{2}$

    (C) $f(x) > 12$
    (D) $f(x) > 0$
    (E) $6 < f(x) < 12$

    handwritten:
    $x^2 + 3x + 2$
    $(x + 1)(x + 2)$

    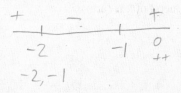

    $-2, -1$

36. If $f(x) = |x| + [x]$, the value of $f(-2.5) + f(1.5)$ is

    (A) −2
    (B) 1
    (C) 1.5
    (D) 2
    (E) 3

USE THIS SPACE FOR SCRATCH WORK

37. If $(\sec x)(\tan x) < 0$, which of the following must be true?

    I. $\tan x < 0$

    II. $\csc x \cot x < 0$

    III. x is in the third or fourth quadrant

    (A) I only
    (B) II only
    (C) III only
    (D) II and III
    (E) I and II

38. At the end of a meeting all participants shook hands with each other. Twenty-eight handshakes were exchanged. How many people were at the meeting?

    (A) 7
    (B) 8
    (C) 14
    (D) 28
    (E) 56

39. Suppose the graph of $f(x) = 2x^2$ is translated 3 units down and 2 units right. If the resulting graph represents the graph of $g(x)$, what is the value of $g(-1.2)$?

    (A) −1.72
    (B) −0.12
    (C) 2.88
    (D) 17.48
    (E) 37.28

40. What is the smallest positive angle x that will make

$$5 - \sin\left(x + \frac{\pi}{6}\right) \text{ a maximum?}$$

    (A) 1.05
    (B) 1.57
    (C) 2.09
    (D) 4.19
    (E) 5.24

41. If $f(x) = ax + b$, which of the following make(s) $f(x) = f^{-1}(x)$?

    I. $a = -1$, $b =$ any real number

    II. $a = 1$, $b = 0$

    III. $a =$ any real number, $b = 0$

    (A) only I
    (B) only II
    (C) only III
    (D) only I and II
    (E) only I and III

USE THIS SPACE FOR SCRATCH WORK

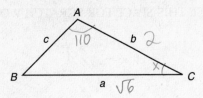

USE THIS SPACE FOR SCRATCH WORK

42. In the figure above, $\angle A = 110°$, $a = \sqrt{6}$, and $b = 2$. What is the value of $\angle C$?

(A) 50°
(B) 25°
(C) 20°
(D) 15°
(E) 10°

$$\frac{\sin 110}{\sqrt{6}} = \frac{\sin x}{2}$$

43. If vector $\vec{v} = \left(1, \sqrt{3}\right)$ and vector $\vec{u} = (3, -2)$, find

the value of $\left|3\vec{v} - \vec{u}\right|$.

(A) 5.4
(B) 6
(C) 7
(D) 7.2
(E) 52

44. A sector of a circle, *AOB*, with a central angle

of $\frac{2\pi}{5}$ and a radius of 5 is bent to form a cone with

vertex at *O*, as shown above. What is the volume of the cone that is formed?

(A) 4.97
(B) 5.13
(C) 6.04
(D) 8.17
(E) 12.31

$$x = \frac{\frac{2\pi}{5}}{2\pi} \cdot 2\pi \cdot 5$$

$$V = \frac{1}{3}\pi r^2 h$$

$$V = \frac{1}{3}\pi 5^2 \sqrt{24}$$

$$2\pi = 2\pi r$$

USE THIS SPACE FOR SCRATCH WORK

45. In $\triangle ABC$ above, $a = 2x$, $b = 3x + 2$, $c = \sqrt{12}$, and $\angle C = 60°$. Find x.

(A) 0.50
(B) 0.64
(C) 0.77
(D) 1.64
(E) 1.78

46. If $\log_a 5 = x$ and $\log_a 7 = y$, then $\log_a \sqrt{1.4} =$

(A) $\frac{1}{2}xy$

(B) $\frac{1}{2}x - y$

(C) $\frac{1}{2}(x + y)$

(D) $\frac{1}{2}(y - x)$

(E) $\frac{y}{2x}$

47. If $f(x) = 3x^2 + 4x + 5$, what must the value of k equal so that the graph of $f(x - k)$ will be symmetric to the y-axis?

(A) -4

(B) $-\dfrac{4}{3}$

(C) $-\dfrac{2}{3}$

(D) $\dfrac{2}{3}$

(E) $\dfrac{4}{3}$

Model Test 1

48. If $f(x) = \cos x$ and $g(x) = 2x + 1$, which of the following is an even function (are even functions)?

    I. $f(x) \cdot g(x)$
    II. $f(g(x))$
    III. $g(f(x))$

    (A) only I
    (B) only II
    (C) only III
    (D) only I and II
    (E) only II and III

49. A cylinder whose base radius is 3 is inscribed in a sphere of radius 5. What is the difference between the volume of the sphere and the volume of the cylinder?

    (A) 88
    (B) 297
    (C) 354
    (D) 448
    (E) 1345

$V_s = \frac{4}{3}\pi r^3 = 523.59$

50. Under which conditions is $\dfrac{xy}{x - y}$ negative?

    (A) $0 < y < x$
    (B) $x < y < 0$
    (C) $x < 0 < y$
    (D) $y < x < 0$
    (E) None of the above

STOP

If there is still time remaining, you may review your answers.

Answer Key

MODEL TEST 1

1. C	14. D	27. C	40. D
2. A	15. A	28. B	41. D
3. A	16. B	29. E	42. C
4. C	17. A	30. C	43. D
5. C	18. E	31. D	44. B
6. B	19. D	32. C	45. B
7. C	20. D	33. C	46. D
8. B	21. B	34. D	47. D
9. E	22. C	35. E	48. C
10. D	23. B	36. D	49. B
11. C	24. A	37. C	50. B
12. C	25. C	38. B	
13. E	26. C	39. D	

ANSWERS EXPLAINED

The following explanations are keyed to the review portions of this book. The number in brackets after each explanation indicates the appropriate section in the Review of Major Topics (Part 2). If a problem can be solved using algebraic techniques alone, [algebra] appears after the explanation, and no reference is given for that problem in the Self-Evaluation Chart at the end of the test.

> In these solutions the following notation is used:
>
> i: calculator unnecessary
> a: calculator helpful or necessary
> g: graphing calculator helpful or necessary

1. i **(C)** Solve for y. $y = \frac{4}{3}x - 4$. Slope $= \frac{4}{3}$. Slope of perpendicular $= -\frac{3}{4}$. [1.2]

2. i **(A)** Range = largest value – smallest value = 18 – 8 = 10. [4.1]

3. i **(A)** Vertical asymptotes occur where the denominator is zero but the numerator is not. The denominator, $x^2 - 49$, factors into $(x + 7)(x - 7)$. Since both numerator and denominator are zero when $x = 7$, a vertical asymptote occurs only at $x = -7$. [1.2]

4. g **(C)** Enter the function f into Y_1 and the function g into Y_2. Evaluate $Y_1(Y_2(2))$ to get the correct answer choice B.
An alternative solution is to evaluate $g(2) = 5$ and $f(5) = \sqrt{13}$, and either use your calculator to evaluate $\sqrt{13}$ or observe that $3 < \sqrt{13} < 4$, indicating 3.61 as the only feasible answer choice. [1.1]

5. g **(C)** Enter the expression into your graphing calculator. [1.4]

6. a **(B)** Complete the square to get $x^2 + (y - 5)^2 = 61$.
Radius $= \sqrt{61}$. $C = 2\pi r = 2\pi\sqrt{61} \approx 49$. [2.1]

7. a **(C)** $\sum\limits_{j=3}^{5} \ln j = \ln 3 + \ln 4 + \ln 5 \approx 4.1$. [3.4]

8. i **(B)** Regardless of what is substituted for x, $f(x)$ still equals 2.

Alternative Solution: $f(x + 2)$ causes the graph of $f(x)$ to be shifted 2 units to the left. Since $f(x) = 2$ for all x, $f(x + 2)$ will also equal 2 for all x. [1.1]

9. g **(E)** Enter the formula for the volume of a sphere $(4/3)\pi x^3$ (in the reference list of formulas) into Y_1. Return to the Home Screen, and enter $Y_1(5) - Y_1(2)$ to get the correct answer choice E.

An alternative solution is to evaluate $V = \frac{4}{3}\pi\ (5^3 - 2^3) = \frac{4}{3}\pi(117)$ directly. [2.2]

10. a **(D)** Since $b^3 = \frac{1}{b^{-3}}$, divide out b^3c^8, leaving $a^5 = 9a^3$. Therefore $a = \pm 3$. [1.4]

11. i (C) $\sin A = \dfrac{8}{10} = \dfrac{4}{5}$. [1.3]

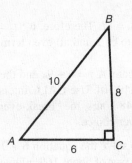

12. i (C) Since the $\dfrac{1}{2}$ power is the square root, $x = \dfrac{1}{2}$ because the square root of 16 is 4.

Since $5^4 = 625$, $x + y = 4$, so that $y = 4 - \dfrac{1}{2} = \dfrac{7}{2}$. [1.4]

13. i (E) $t = \dfrac{y}{2}$. Eliminate the parameter and get $x = \dfrac{y^2}{4} + 1$ or $y^2 = 4x - 4$. [1.6]

14. i (D) $f(1) = 0$ and $f(2) = 0$ imply that $x - 1$ and $x - 2$ are factors of $f(x)$. Their product, $x^2 - 3x + 2$, is also a factor. [1.2]

15. g (A) With the calculator in either mode, evaluate $\sin(\tan^{-1}(1/3))$ to get the correct answer choice A. An alternative solution is to define $\theta = \tan^{-1}\left(\dfrac{1}{3}\right)$, so that

$\tan \theta = \dfrac{1}{3}$. Since θ is in Quadrant I, use a right triangle, the Pythagorean Theorem, and

the definition to find $\sin \theta = \dfrac{1}{\sqrt{10}} \approx 0.32$. [1.3]

16. i (B) $a^2 = z^2 \cos^2 \theta$ and $b^2 = z^2 \sin^2 \theta$, so $a^2 + b^2 = z^2(\cos^2 \theta + \sin^2 \theta) = z^2$ because

$\cos^2 \theta + \sin^2 \theta = 1$. Since $\sqrt{z^2} = z$ when $z > 0$, the correct answer choice is B. [1.3]

17. i (A) Sketch a graph of the three vertices. The base is $|u|$ and the altitude is 8. Therefore, the area is $4|u|$. [2.1]

18. g (E) Plot the graph of $y = \dfrac{5}{x - 2}$ in the standard window, and observe the asymptote

at $x = 2$. This says that no value of k can make $f(x)$ continuous at $x = 2$.

An alternative solution is to observe that $x = 2$ makes the denominator of $f(x)$ equal to zero, thereby implying that $x = 2$ is a vertical asymptote. Thus, $f(x)$ cannot be made continuous at the point with that x value. [1.6]

19. i (D) There are 5 prime numbers less than 13: 2, 3, 5, 7, 11. Three of these are less than

7, so the correct probability is $\dfrac{3}{5}$. [4.2]

20. a **(D)** Substituting for x^2 and solving for y gives $4(9 - y^2) + 8y^2 = 64$. $4y^2 = 28$, and so $y^2 = 7$ and $y = \pm\sqrt{7}$. [2.1]

21. i **(B)** $t_n \cdot t_{n+1} = K$. $2 \cdot 6 = K = 12$. Therefore, $6 \cdot t_3 = 12$, and so $t_3 = 2$. Continuing this process gives all odd terms to be 2 and all even terms to be 6. [3.4]

22. g **(C)** Graph the cost of Company A $y = 6 + 9x$ and the cost of Company B $y = 25 + 2.25x$ in a window $x\varepsilon[0,10]$ and $y\varepsilon[0,50]$. Use CALC/intersect to find the x-coordinate of the point of intersection at 2.8148 hours, the "break-even" point. Multiply by 60 to convert this time to the correct answer choice.

An alternative solution is to solve the equation $6 + 9x = 25 + 2.25x$ and multiply the solution by 60 to get the answer of about 169 minutes. [1.2]

23. i **(B)** The probability that both teams will win is pq. The probability that both will lose is $(1 - p)(1 - q)$. The probability that only one will win is $1 - [pq + (1 - p)(1 - q)] = 1 - (pq + 1 - p - q + pq) = p + q - 2pq$.

Alternative Solution: The probability that the Giants will win and the Raiders will lose is $p(1 - q)$. The probability that the Raiders will win and the Giants will lose is $q(1 - p)$. Therefore, the probability that either one of these results will occur is $p(1 - q) + q(1 - p) = p + q - 2pq$. [4.2]

24. g **(A)** Calculate the common ratio as $\dfrac{1}{1/3} = 3$. The first term is $\dfrac{1}{3}$ so the nth term is

$t_n = \left(\dfrac{1}{3}\right)3^{n-1}$. Use the sum and sequence features of your calculator to evaluate the sum of the first 10 terms in the generated sequence:

$$\text{LIST/MATH/sum(LIST/OPS/}$$
$$\text{seq}((1/3)3\verb|^|X, X, 0.9)) = 9841.333\cdots$$

The range is 0 to 9 instead of 1 to 10 because the formula for t_n uses the exponent $n - 1$.

An alternative solution is to use the formula for the sum of a geometric series:

$$S_n = \frac{t_1(1 - r^n)}{1 - r} = \frac{(1/3)(1 - 3^{10})}{1 - 3} = 9841\frac{1}{3}$$

[3.4]

25. g **(C)** No calculator currently on the market can compute 453!, so doing this problem requires some knowledge of factorial arithmetic. The easiest solution to the problem is to observe that $\dfrac{453!}{450!3!}$ is the number of combinations of 453 taken 3 at a time $(_{453}C_3)$. Enter 453MATH/PRB/nCr3 into your calculator to find that the correct answer choice is C.

An alternative solution is to simplify $\dfrac{453!}{450!3!}$ to $\dfrac{453 \cdot 452 \cdot 451}{3 \cdot 2 \cdot 1} = 1.5\cdots\times 10^7$. [3.1]

26. a (C) Solve for y: $y = \dfrac{3}{2}x + \dfrac{7}{2}$. Slope $= \dfrac{\Delta y}{\Delta x} = \dfrac{3}{2}$. Tan A also equals $\dfrac{\Delta y}{\Delta x}$. Therefore,

$\tan A = \dfrac{3}{2}$. $\text{Tan}^{-1}\left(\dfrac{3}{2}\right) = \angle A \approx 56°$. [1.3]

27. g (C) Put your calculator in radian mode and graph $y = 3\sin(2(x + 2\pi/3))$ using Ztrig. Since the answer choices are pretty far apart, you can determine the correct answer choice by tracing to the first positive x-intercept 1.05.

An alternative solution can be found by solving $3\sin(2(x + 2\pi/3)) = 0$. This means that $x + 2\pi/3 = 0, \pi, \ldots$. The equation $x + 2\pi/3 = \pi$ yields the smallest positive solution $x = \dfrac{\pi}{3} \approx 1.05$. [1.3]

28. i (B) Divide the equation through by 16 to get $\dfrac{(x-4)^2}{16} + \dfrac{(y-3)^2}{4} = 1$. This is the equation of an ellipse with $a^2 = 16$. The sum of the distances to the foci $= 2a = 8$. [2.1]

29. a (E) If the roots are r and $2r$, their sum $= -\dfrac{b}{a} = 3r = -\dfrac{k}{1}$ and their product $= \dfrac{c}{a} =$

$2r^2 = \dfrac{54}{1}$. Therefore, $r = \pm\sqrt{27}$ and $k = \pm3\sqrt{27} \approx \pm15.6$.

Alternative Solution: If the roots are r and $2r$, $(x - r)(x - 2r) = 0$. Multiply to obtain $x^2 - 3r + 2r^2 = 0$, which represents $x^2 + kx + 54 = 0$. Thus, $-3r = k$ and $2r^2 = 54$.

Since $r = -\dfrac{k}{3}$, then $2\left(-\dfrac{k}{3}\right) = 54$ and $k = \pm3\sqrt{27} \approx 15.6$. [1.2]

30. i (C) Substituting -1 for x gives 3.

Alternative Solution: Use synthetic division to get

$$
\begin{array}{r|rrrrr}
-1 & 3 & 7 & 8 & -2 & -3 \\
   &   & -3 & -4 & -4 & 6 \\
\hline
   & 3 & 4 & 4 & -6 & \boxed{3} = \text{remainder} \; [1.2]
\end{array}
$$

31. a (D) The inverse of $f(x) = e^x$ is $f^{-1}(x) = \ln x$. $g(2) = e^2 + \ln 2 \approx 8.1$. [1.4]

32. g (C) This is a recursively defined sequence. Press 3 (x_0) ENTER on your calculator. Then enter $\sqrt{4 + \text{Ans}}$ and press ENTER 3 times to get $x_3 \approx 2.56$. [3.4]

33. i (C) If the graph passes through the origin, $x = 0$ and $y = 0$, then $\dfrac{4k^2}{1} - \dfrac{9k^2}{3} = 1$. $k^2 = 1$, and so $k = \pm1$. [2.1]

34. g (D) Graph $y = \dfrac{1 - \cos x}{\sin x}$ and $y = \dfrac{\sqrt{3}}{3}$ using Ztrig in degree mode. Find the point of intersection with CALC/intersect to arrive at the correct answer choice D.

An alternative solution uses the identities $\tan \dfrac{\theta}{2} = \dfrac{1 - \cos x}{\sin x}$ and $\tan 30° = \dfrac{\sqrt{3}}{3}$ to deduce $\dfrac{\theta}{2} = 30°$, so $\theta = 60°$. [1.3]

35. g (E) The problem is asking for the range of $f(x)$ values for values of x that satisfy the inequality. First graph the inequality in Y_1, starting with the standard window and zooming in until the x values for the portion of the graph that falls below the x-axis can be identified as the interval $(-2, -1)$. Then enter the formula for $f(x)$ in Y_2. Although it can be done graphically, the simplest way to find the range of values of $f(x)$ that correspond to $x \varepsilon (-2, -1)$ is to use the TABLE function. Deselect Y_1 and enter TBLSET and set TblStart to -2, $\Delta Tbl = 0.1$, and Indpnt and Depend to Auto. Then enter TABLE and observe that the Y_2 values range from 12 to 6 as x ranges from -2 to -1, yielding the correct answer choice D.

An alternative solution is to solve the inequality algebraically by solving the associated equation $x^2 + 3x + 2 = 0$ and testing points. The left side of the equation factors as $(x + 2)(x + 1)$, and the Zero Product Property implies that $x = -2$ or $x = -1$. Points inside the interval $(-2, -1)$ satisfy the inequality, while those outside it do not. Since the graph of $f(x)$ is a parabola and $f(-2) = 12$ and $f(-1) = 6$, $f(x)$ takes the range of values between 12 and 6. [1.2]

36. g (D) Recall that the notation $[x]$ means the greatest integer less than or equal to x. Enter abs$(x) +$ int(x) into Y_1. Return to the Home Screen, and enter $Y_1(-2.5) + Y_1(1.5)$ to get the correct answer choice D.

An alternative solution evaluates $|-2.5| + [-2.5] + |1.5| + [1.5]$ without the aid of a calculator. Of these 4 values, only $[-2.5]$ is tricky since $[-2.5] = -3$, not -2. Thus, $|-2.5| + [-2.5] + |1.5| + [1.5] = 2.5 - 3 + 1.5 + 1 = 2$. [1.6]

37. i (C) Set up the following table.

	Q1	Q2	Q3	Q4
$\sec x$	+	−	−	+
$\tan x$	+	−	+	−
$\cot x$	+	−	+	−
$\csc x$	+	+	−	−

The product $\sec x \tan x$ is negative only when its factors have different signs, so III is the only true statement. [1.3]

38. i (B) $\dbinom{x \text{ people}}{2} = 28$. $\dfrac{x(x-1)}{2 \cdot 1} = 28$. $x^2 - x = 56$. $x = 8$. [3.1]

39. a (D) Since the function g is f translated 3 down and 2 right, $g(x) = f(x - 2) - 3$. Therefore, $g(-1.2) = f(-3.2) - 3 = 2(-3.2)^2 - 3 = 17.48$. [2.1]

40. g (D) Set your calculator to radian mode and graph the function in an $x \varepsilon [0, 2\pi]$ and $y \varepsilon [3, 7]$ window. Use CALC/maximum to find 4.19 as the smallest positive x where a maximum occurs.

An alternative solution is to observe that the function will be a maximum when $\sin\left(x + \dfrac{\pi}{6}\right) = -1$. This will happen when $x + \dfrac{\pi}{6} = \dfrac{3\pi}{2}$, or when $x = \dfrac{4\pi}{3} \approx 4.19$. [1.3]

41. a **(D)** The graph of f must be symmetric about the line $y = x$. In other words, interchanging x and y must leave the graph unchanged. In I, $x = -y + b$, which is equivalent to $y = -x + b$, which is symmetric about $y = x$. In II, $x = y$. In III, $x = ay$, or $y = \dfrac{x}{a}$ [1.1]

42. a **(C)** Law of sines:

$$\frac{\sin 110°}{\sqrt{6}} = \frac{\sin B}{2}; \sin B = \frac{2\sin 110°}{\sqrt{6}} \approx 0.7673. \; \text{Sin}^{-1}(0.7673) = \angle B = 50°. \text{ Therefore,}$$

$\angle C = 180° - 110° - 50° = 20°.$ [1.3]

43. a **(D)** $\left|3\vec{v} - \vec{u}\right| = \left|\left(3, 3\sqrt{3}\right) - (3, -2)\right| = \left|\left(0, 3\sqrt{3} + 2\right)\right| = \sqrt{0^2 + \left(3\sqrt{3} + 2\right)^2} = 3\sqrt{3} + 2 \approx 7.2.$

[3.5]

44. a **(B)** Circumference of base of cone = length of arc $AB = 5\left(\dfrac{2\pi}{5}\right) = 2\pi.$

Circumference $= 2\pi r = 2\pi.$ Therefore, radius of base = 1. By the Pythagorean

Theorem, height of cone $= \sqrt{24}$. Therefore, volume $= \dfrac{1}{3}\pi r^2 h = \dfrac{1}{3}\pi \cdot 1 \cdot \sqrt{24} \approx 5.13.$
[2.2]

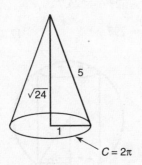

45. g **(B)** Law of Cosines:

$$12 = (2x)^2 + (3x + 2)^2 - 2 \cdot 2x \cdot (3x + 2)\cos 60°.$$
$$12 = 4x^2 + 9x^2 + 12x + 4 - \left(12x^2 + 8x\right) \cdot \frac{1}{2}.$$

$7x^2 + 8x - 8 = 0.$

Use program QUADFORM to get $x = \pm 0.64.$

Since a side of a triangle must be positive, x can equal only 0.64. [1.3]

46. i **(D)** $\log_a \sqrt{1.4} = \log_a \left(\dfrac{7}{5}\right)^{1/2} = \dfrac{1}{2}\left(\log_a 7 - \log_a 5\right) = \dfrac{1}{2}(y - x).$ [1.4]

47. g (D) Graph $y = 3x^2 + 4x + 5$ in the standard window, and observe that the graph must be moved slightly to the right to be symmetric to the y-axis. Therefore, k must be positive. Use CALC/minimum to find the vertex of the parabola and observe that its

x coordinate is –0.66666. . . . If the function entered into Y_1, set $Y_2 = Y_1\left(x - \dfrac{2}{3}\right)$

and graph Y_2 to verify this answer. [2.1]

48. g (C) Use ZTrig to plot the graphs of $y = (\cos x) \cdot (2x + 1)$, $y = \cos(2x + 1)$, and $y = 2(\cos x) + 1$ to see that only the third graph is symmetric about the y-axis and thus represents an even function.

An alternative solution is to use your knowledge of transformations. Although f is an even function, g is not; therefore, (I) $f \cdot g$ is not even. Also, $f(g(x)) = \cos(2x + 1)$, which is a cosine curve shifted less than π to the left. Thus, $f(g(x))$ (II) is not even. However, $g(f(x)) = 2 \cos x + 1$ is a cosine curve with period 2π, amplitude 2, shifted 1 unit up. Thus, $g(f(x))$ (III) is even. [1.1]

49. a (B) Height of cylinder is 8.

Volume of sphere $= \dfrac{4}{3}\pi r^3 = \dfrac{4}{3}\pi(125) = \dfrac{500\pi}{3}$.

Volume of cylinder $= \pi r^2 h = \pi(9)8$.

Difference $= \dfrac{500}{3}\pi - 72\pi \approx 523.6 - 226.7$

$\approx 297.$ [2.2]

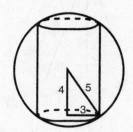

50. i (B) In answer choice B, x and y have the same sign, and x is less than y. Therefore xy is positive, $x - y$ is negative, and the quotient is negative. The numerators and denominators in answer choices A, C, and D both have the same sign, so the quotients are positive. [algebra]

Self-Evaluation Chart for Model Test 1

Subject Area	Questions and Review Section						Right	Number Wrong	Omitted
Algebra and Functions (20 questions)	1	3 ^w	4	5	8	10	4	1	0
	1.2	1.2	1.1	1.4	1.1	1.4			
	12	13	14	18	22	29 ^o	5	0	1
	1.4	1.6	1.2	1.6	1.2	1.2			
	30	31	35 ^o	36			3	0	1
	1.2	1.4	1.2	1.6					
	41	46 ^o	48 ^o	50 ^o			1	0	3
	1.1	1.4	1.1						
Trigonometry (10 questions)	11	15	16	26	27	34	6	0	0
	1.3	1.3	1.3	1.3	1.3	1.3			
	37 ^w	40	42 ^w	45			2	2	0
	1.3	1.3	1.3	1.3					
Coordinate and Three-Dimensional Geometry (10 questions)	6	9	17	20	28	33	6	0	0
	2.1	2.2	2.1	2.1	2.1	2.1			
	39	44 ^o	47 ^o	49 ^o			1	0	3
	2.1	2.2	2.1	2.2					
Numbers and Operations (7 questions)	7	21 ^w	24	25	32	38 ^o	43 ^o	4	1	2
	3.4	3.4	3.4	3.1	3.4	3.1	3.5			
Data Analysis, Statistics, and Probability (3 questions)	2	19 ^o	23				2	0	1
	4.1	4.2	4.2						
TOTALS							34	4	11

Evaluate Your Performance Model Test 1

Rating	Number Right
Excellent	41–50
Very good	33–40
Above average	25–32
Average	15–24
Below average	Below 15

Calculating Your Score

Raw score R = number right $- \dfrac{1}{4}$ (number wrong), rounded = _____ 33

Approximate scaled score $S = 800 - 10(44 - R) =$ _____ 690

If $R \geq 44$, $S = 800$.

Answer Sheet
MODEL TEST 2

1 Ⓐ Ⓑ Ⓒ Ⓓ   14 Ⓐ Ⓑ Ⓒ Ⓓ   27 Ⓐ Ⓑ Ⓒ Ⓓ   40 Ⓐ Ⓑ Ⓒ Ⓓ

2 Ⓐ Ⓑ Ⓒ Ⓓ   15 Ⓐ Ⓑ Ⓒ Ⓓ   28 Ⓐ Ⓑ Ⓒ Ⓓ   41 Ⓐ Ⓑ Ⓒ Ⓓ

3 Ⓐ Ⓑ Ⓒ Ⓓ   16 Ⓐ Ⓑ Ⓒ Ⓓ   29 Ⓐ Ⓑ Ⓒ Ⓓ   42 Ⓐ Ⓑ Ⓒ Ⓓ

4 Ⓐ Ⓑ Ⓒ Ⓓ   17 Ⓐ Ⓑ Ⓒ Ⓓ   30 Ⓐ Ⓑ Ⓒ Ⓓ   43 Ⓐ Ⓑ Ⓒ Ⓓ

5 Ⓐ Ⓑ Ⓒ Ⓓ   18 Ⓐ Ⓑ Ⓒ Ⓓ   31 Ⓐ Ⓑ Ⓒ Ⓓ   44 Ⓐ Ⓑ Ⓒ Ⓓ

6 Ⓐ Ⓑ Ⓒ Ⓓ   19 Ⓐ Ⓑ Ⓒ Ⓓ   32 Ⓐ Ⓑ Ⓒ Ⓓ   45 Ⓐ Ⓑ Ⓒ Ⓓ

7 Ⓐ Ⓑ Ⓒ Ⓓ   20 Ⓐ Ⓑ Ⓒ Ⓓ   33 Ⓐ Ⓑ Ⓒ Ⓓ   46 Ⓐ Ⓑ Ⓒ Ⓓ

8 Ⓐ Ⓑ Ⓒ Ⓓ   21 Ⓐ Ⓑ Ⓒ Ⓓ   34 Ⓐ Ⓑ Ⓒ Ⓓ   47 Ⓐ Ⓑ Ⓒ Ⓓ

9 Ⓐ Ⓑ Ⓒ Ⓓ   22 Ⓐ Ⓑ Ⓒ Ⓓ   35 Ⓐ Ⓑ Ⓒ Ⓓ   48 Ⓐ Ⓑ Ⓒ Ⓓ

10 Ⓐ Ⓑ Ⓒ Ⓓ   23 Ⓐ Ⓑ Ⓒ Ⓓ   36 Ⓐ Ⓑ Ⓒ Ⓓ   49 Ⓐ Ⓑ Ⓒ Ⓓ

11 Ⓐ Ⓑ Ⓒ Ⓓ   24 Ⓐ Ⓑ Ⓒ Ⓓ   37 Ⓐ Ⓑ Ⓒ Ⓓ   50 Ⓐ Ⓑ Ⓒ Ⓓ

12 Ⓐ Ⓑ Ⓒ Ⓓ   25 Ⓐ Ⓑ Ⓒ Ⓓ   38 Ⓐ Ⓑ Ⓒ Ⓓ

13 Ⓐ Ⓑ Ⓒ Ⓓ   26 Ⓐ Ⓑ Ⓒ Ⓓ   39 Ⓐ Ⓑ Ⓒ Ⓓ

Model Test 2

T ear out the preceding answer sheet. Decide which is the best choice by rounding your answer when appropriate. Blacken the corresponding space on the answer sheet. When finished, check your answers with those at the end of the test. For questions that you got wrong, note the sections containing the material that you must review. Also if you do not fully understand how you arrived at some of the correct answers, you should review the appropriate sections. Finally, fill out the self-evaluation chart on page 252 in order to pinpoint the topics that give you the most difficulty.

50 questions: 1 hour

Directions: Decide which answer choice is best. If the exact numerical value is not one of the answer choices, select the closest approximation. Fill in the oval on the answer sheet that corresponds to your choice.

Notes:
(1) You will need to use a scientific or graphing calculator to answer some of the questions.
(2) You will have to decide whether to put your calculator in degree or radian mode for some problems.
(3) All figures that accompany problems are plane figures unless otherwise stated. Figures are drawn as accurately as possible to provide useful information for solving the problem, except when it is stated in a particular problem that the figure is not drawn to scale.
(4) Unless otherwise indicated, the domain of a function is the set of all real numbers for which the functional value is also a real number.

Model Test 2

Reference Information. The following formulas are provided for your information.

Volume of a right circular cone with radius r and height h: $V = \dfrac{1}{3}\pi r^2 h$

Lateral area of a right circular cone if the base has circumference c and slant height is l:

$S = \dfrac{1}{2} cl$

Volume of a sphere of radius r: $V = \dfrac{4}{3}\pi r^3$

Surface area of a sphere of radius r: $S = 4\pi r^2$

Volume of a pyramid of base area B and height h: $V = \dfrac{1}{3} Bh$

1. If $f(x) = \dfrac{x-2}{x^2-4}$, for what value(s) of x does the graph of $f(x)$ have a vertical asymptote?

   (A) $-2, 0,$ and 2
   (B) -2 and 2
   (C) 2
   (D) 0
   (E) -2

2. If a regular square pyramid has x pairs of parallel edges, then x equals

   (A) 1
   (B) 2
   (C) 4
   (D) 8
   (E) 12

3. Log $(a^2 - b^2) =$

   (A) $\log a^2 - \log b^2$
   (B) $\log \dfrac{a^2}{b^2}$
   (C) $\log \dfrac{a+b}{a-b}$
   (D) $2 \cdot \log a - 2 \cdot \log b$
   (E) $\log(a+b) + \log(a-b)$

4. The sum of the roots of the equation
   $\left(x - \sqrt{2}\right)^2 \left(x + \sqrt{3}\right)\left(x - \sqrt{5}\right) = 0$ is

   (A) 1.9
   (B) 2.2
   (C) 2.5
   (D) 3.3
   (E) 6.8

5. If the graph of $x + 2y + 3 = 0$ is perpendicular to the graph of $ax + 3y + 2 = 0$, then a equals

   (A) -6
   (B) $-\dfrac{3}{2}$
   (C) $\dfrac{2}{3}$
   (D) $\dfrac{3}{2}$
   (E) 6

6. The maximum value of $6 \sin x \cos x$ is

   (A) $\dfrac{1}{3}$

   (B) 1
   (C) 2.6
   (D) 3
   (E) 6

7. If $f(r, \theta) = r \cos \theta$, then $f(2, 3) =$

   (A) -3.00
   (B) -1.98
   (C) 0.10
   (D) 1.25
   (E) 2.00

8. If 5 and -1 are both zeros of the polynomial $P(x)$, then a factor of $P(x)$ is

   (A) $x^2 - 5$
   (B) $x^2 - 4x + 5$
   (C) $x^2 + 4x - 5$
   (D) $x^2 + 5$
   (E) $x^2 - 4x - 5$

9. $i^{14} + i^{15} + i^{16} + i^{17} =$

   (A) 0
   (B) 1
   (C) $2i$
   (D) $1 - i$
   (E) $2 + 2i$

10. When the graph of $y = \sin 2x$ is drawn for all values of x between $10°$ and $350°$, it crosses the x-axis

    (A) zero times
    (B) one time
    (C) two times
    (D) three times
    (E) six times

11. The third term of an arithmetic sequence is 15, and the seventh term is 23. What is the first term?

    (A) 1
    (B) 6
    (C) 9
    (D) 11
    (E) 13

USE THIS SPACE FOR SCRATCH WORK

12. A particular sphere has the property that its surface area has the same numerical value as its volume. What is the length of the radius of this sphere?

(A) 1
(B) 2
(C) 3
(D) 4
(E) 6

$\frac{4}{3}\pi r^3 = 4\pi r^2$

$\frac{4}{3}r = 4$

$r = 3$

13. Suppose $\sin x = -\dfrac{3}{5}$, $\dfrac{3\pi}{2} < x < 2\pi$. Then $\sin 2x =$

(A) $-\dfrac{6}{5}$

(B) $-\dfrac{24}{25}$

(C) $-\dfrac{12}{25}$

(D) $\dfrac{9}{25}$

(E) $\dfrac{24}{25}$

14. The pendulum on a clock swings through an angle of 1 radian, and the tip sweeps out an arc of 12 inches. How long is the pendulum?

(A) 3.8 inches
(B) 6 inches
(C) 7.6 inches
(D) 12 inches
(E) 35 inches

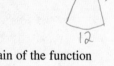

$12 = \dfrac{1}{2\pi} \cdot 2\pi r$

15. What is the domain of the function

$f(x) = 4 - \sqrt{3x^3 - 7}$?

(A) $x \geq 1.33$
(B) $x \geq 1.53$
(C) $x \geq 2.33$
(D) $x \leq -1.33$ or $x \geq 1.33$
(E) $x \leq -2.33$ or $x \geq 2.33$

$3x^3 - 7 \geq 0$

$x^3 \geq \dfrac{7}{3}$

16. If $x + y = 90°$, which of the following must be true?

(A) $\cos x = \cos y$
(B) $\sin x = -\sin y$
(C) $\tan x = \cot y$
(D) $\sin x + \cos y = 1$
(E) $\tan x + \cot y = 1$

17. The graph of the equation $y = x^3 + 5x + 1$

   (A) does not intersect the *x*-axis
   (B) intersects the *x*-axis at one and only one point
   (C) intersects the *x*-axis at exactly three points
   (D) intersects the *x*-axis at more than three points
   (E) intersects the *x*-axis at exactly two points

18. The length of the radius of the sphere
   $x^2 + y^2 + z^2 + 2x - 4y = 10$ is

   (A) 3.16
   (B) 3.38
   (C) 3.46
   (D) 3.74
   (E) 3.87

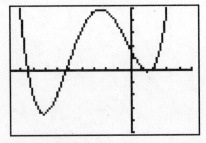

```
WINDOW
  Xmin=■10
  Xmax=5
  Xscl=1
  Ymin=-250
  Ymax=250
  Yscl=50
  Xres=1
```

19. The graph of $y = x^4 + 11x^3 + 9x^2 - 97x + c$ is shown above with the window shown next to it. Which of the following values could be *c*?

   (A) –2820
   (B) –80
   (C) 80
   (D) 250
   (E) 2820

20. Which of the following is the solution set for
   $x(x - 3)(x + 2) > 0$?

   (A) $x < -2$
   (B) $-2 < x < 3$
   (C) $-2 < x < 3$ or $x > 3$
   (D) $x < -2$ or $0 < x < 3$
   (E) $-2 < x < 0$ or $x > 3$

Model Test 2

21. Which of the following is the equation of the circle that has its center at the origin and is tangent to the line with equation $3x - 4y = 10$?

    (A) $x^2 + y^2 = 2$
    (B) $x^2 + y^2 = 4$
    (C) $x^2 + y^2 = 3$
    (D) $x^2 + y^2 = 5$
    (E) $x^2 + y^2 = 10$

$-4y = -3x + 10$

$y = \frac{3}{4}x - \frac{10}{4}$

22. If $f(x) = 3 - 2x + x^2$, then $\dfrac{f(x+t) - f(x)}{t} =$

    (A) $t^2 + 2xt - 2t$
    (B) $x^2t^2 - 2xt + 3$
    (C) $t + 2x - 2$
    (D) $2x - 2$
    (E) none of the above

$\dfrac{3 - 2(x+t) + (x+t)^2 - (3 - 2x + x^2)}{t} = \dfrac{\begin{array}{l}3 - 2x - 2t + x^2 + t^2 + 2xt \\ -3 + 2x - x^2\end{array}}{t}$

$= \dfrac{-2t + t^2 + 2xt}{}$

$= \dfrac{t(-2 + t + 2x)}{t}$

23. If $f(x) = x^3$ and $g(x) = x^2 + 1$, which of the following is an odd function (are odd functions)?

    I. $f(x) \cdot g(x)$
    II. $f(g(x))$ ✓
    III. $g(f(x))$ ✓

    (A) only I
    (B) only II
    (C) only III
    (D) only II and III
    (E) I, II, and III

24. In how many ways can a committee of four be selected from nine men so as to always include a particular man?

    (A) 48
    (B) 56
    (C) 70
    (D) 84
    (E) 126

$\underline{1} \quad \underline{8} \quad \underline{7} \quad \underline{6}$

25. When $f(x) = \sin x$ and $g(x) = \cos x$ over the interval $(0, 2\pi)$, and $M(f, g)$ is defined to be the maximum of f and g, the graph of $M(f, g)$ looks like which one of the following?

(A)

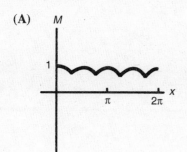

(B)

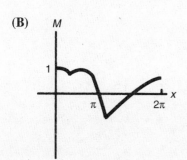

(C)

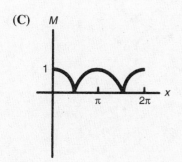

(D)

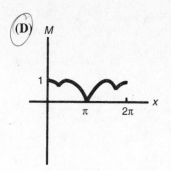

(E)

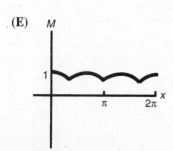

26. If the mean of the set of data 1, 2, 3, 1, 2, 5, *x* is $3.\overline{27}$ what is the value of *x*?

(A) –10.7
(B) 2.5
(C) 5.6
(D) 7.4
(E) 8.9

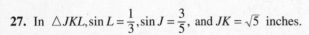

27. In $\triangle JKL, \sin L = \frac{1}{3}, \sin J = \frac{3}{5}$, and $JK = \sqrt{5}$ inches.

The length of *KL*, in inches, is

(A) 1.7
(B) 3.0
(C) 3.5
(D) 3.9
(E) 4.0

28. Matrix *X* has *r* rows and *c* columns, and matrix *Y* has *c* rows and *d* columns, where *r*, *c*, and *d* are different. Which of the following statements must be false?

  I. The product *YX* exists
  II. The product of *XY* exists and has *r* rows and *d* columns.
  III. The product *XY* exists and has *c* rows and *c* columns.

(A) I only
(B) II only
(C) III only
(D) I and II
(E) I and III

29. Which of the following statements is logically equivalent to: "If he studies, he will pass the course."

(A) He passed the course; therefore, he studied.
(B) He did not study; therefore, he will not pass the course.
(C) He did not pass the course; therefore he did not study.
(D) He will pass the course only if he studies.
(E) None of the above.

30. If $f(x) = x - 7$ and $g(x) = \sqrt{x}$, what is the domain of $g \circ f$?

(A) $x \le 0$
(B) $x \ge -7$
(C) $x \ge 0$
(D) $x \ge 7$
(E) all real numbers

31. In $\triangle ABC$, $a = 1$, $b = 4$, and $\angle C = 30°$. The length of c is

(A) 4.6
(B) 3.6
(C) 3.2
(D) 2.9
(E) 2.3

32. The solution set of $3x + 4y < 0$ lies in which quadrants?

(A) I only
(B) I and II
(C) I, II, and III
(D) II, III, and IV
(E) I, II, III, and IV

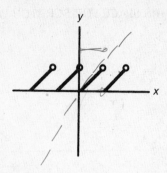

33. Which of the following could represent the inverse of the function graphed above?

(A)
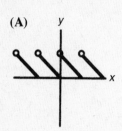

(B)
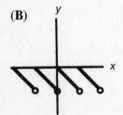

(C)
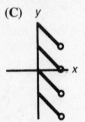

(D)
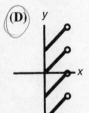

(E)
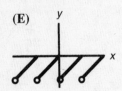

34. If f is a linear function and $f(-2) = 11$, $f(5) = -2$, and $f(x) = 4.3$, what is the value of x?

(A) –3.1
(B) –1.9
(C) 1.6
(D) 2.9
(E) 3.2

35. A taxicab company wanted to determine the fuel cost of its fleet. A sample of 30 vehicles was selected, and the fuel cost for the last month was tabulated for each vehicle. Later it was discovered that the highest amount was mistakenly recorded with an extra zero, so it was 10 times the actual amount. When the correction was made, this was still the highest amount. Which of the following remained the same after the correction was made?

(A) mean
(B) median
(C) mode
(D) range
(E) standard deviation

36. The range of the function $y = x^{-2/3}$ is

(A) $y < 0$
(B) $y > 0$
(C) $y \geq 0$
(D) $y \leq 0$
(E) all real numbers

37. If $\cos x = \dfrac{4}{5}$ and $\dfrac{3\pi}{2} \leq x \leq 2\pi$, then $\tan 2x =$

(A) $-\dfrac{24}{7}$

(B) $-\dfrac{24}{25}$

(C) $-\dfrac{3}{4}$

(D) $-\dfrac{7}{24}$

(E) $\dfrac{7}{25}$

USE THIS SPACE FOR SCRATCH WORK

38. A coin is tossed three times. Given that at least one head appears, what is the probability that exactly two heads will appear?

(A) $\dfrac{3}{8}$

(B) $\dfrac{3}{7}$

(C) $\dfrac{5}{8}$

(D) $\dfrac{3}{4}$

(E) $\dfrac{7}{8}$

39. A unit vector parallel to vector $\vec{V} = (2, -3, 6)$ is vector

(A) $(-2, 3, -6)$
(B) $(6, -3, 2)$
(C) $(-0.29, 0.43, -0.86)$
(D) $(0.29, 0.43, -0.86)$
(E) $(-0.36, -0.54, 1.08)$

40. What is the equation of the horizontal asymptote of

the function $f(x) = \dfrac{(2x-1)(x+3)}{(x+3)^2}$

(A) $y = -9$
(B) $y = -3$
(C) $y = 0$
(D) $y = \dfrac{1}{2}$
(E) $y = 2$

41. If $f(x) = \dfrac{x}{x-1}$ and $f^2(x) = f(f(x))$, $f^3(x) = f(f^2(x))$,

. . . , $f^n(x) = f(f^{n-1}(x))$, where n is a positive integer greater than 1, what is the smallest value of n such that $f^n(x) = f(x)$?

(A) 2
(B) 3
(C) 4
(D) 6
(E) No value of n works.

42. A committee of 5 people is to be selected from 6 men and 9 women. If the selection is made randomly, what is the probability that the committee consists of 3 men and 2 women?

(A) $\frac{1}{9}$

(B) $\frac{240}{1001}$

(C) $\frac{1}{3}$

(D) $\frac{1260}{3003}$

(E) $\frac{13}{18}$

43. Three consecutive terms, in order, of an arithmetic sequence are $x+\sqrt{2}$, $2x+\sqrt{3}$, and $5x-\sqrt{5}$. Then x equals

(A) 2.14
(B) 2.45
(C) 2.46
(D) 3.24
(E) 3.56

44. The graph of $xy - 4x - 2y - 4 = 0$ can be expressed as a set of parametric equations. If $y = \dfrac{4t}{t-3}$ and $x = f(t)$, then $f(t) =$

(A) $t+1$
(B) $t-1$
(C) $3t-3$
(D) $\dfrac{t-3}{4t}$
(E) $\dfrac{t-3}{2}$

45. If $f(x) = ax^2 + bx + c$, how must a and b be related so that the graph of $f(x-3)$ will be symmetric about the y-axis?

(A) $a = b$
(B) $b = 0$, a is any real number
(C) $b = 3a$
(D) $b = 6a$
(E) $a = \dfrac{1}{9}b$

46. The graph of $y = \log_5 x$ and $y = \ln 0.5x$ intersect at a point where x equals

(A) 6.24
(B) 5.44
(C) 1.69
(D) 1.14
(E) 1.05

47. What is the value of x if $\pi \leq x \leq \dfrac{3\pi}{2}$ and $\sin x = 5 \cos x$?

$0 = 5\cos x - \sin x$

(A) 3.399
(B) 3.625
(C) 4.515
(D) 4.623
(E) 4.663

48. The area of the region enclosed by the graph of the polar curve $r = \dfrac{1}{\sin\theta + \cos\theta}$ and the x- and y-axes is

$\dfrac{1}{2} \cdot 1 \cdot 1$

(A) 0.48
(B) 0.50
(C) 0.52
(D) 0.98
(E) 1.00

49. A rectangular box has dimensions of length = 6, width = 4, and height = 5. The measure of the angle formed by a diagonal of the box with the base of the box is

(A) 27°
(B) 35°
(C) 40°
(D) 44°
(E) 55°

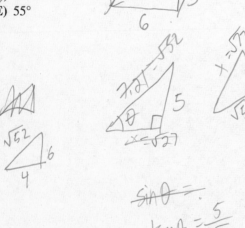

$\sin\theta =$

$\tan\theta = \dfrac{5}{\sqrt{27}}$

50. If (x,y) represents a point on the graph of $y = 2x + 1$, which of the following could be a portion of the graph of the set of points (x,y^2)?

(A)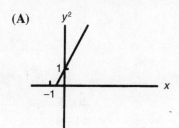

(B)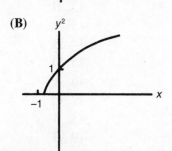

(C)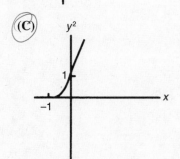

(D)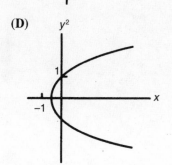

(E)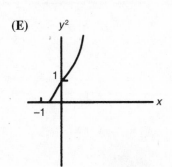

STOP

Answer Key

MODEL TEST 2

1. **E**	14. **D**	27. **E**	40. **E**
2. **B**	15. **A**	28. **B**	41. **B**
3. **E**	16. **C**	29. **C**	42. **B**
4. **D**	17. **B**	30. **D**	43. **A**
5. **A**	18. **E**	31. **C**	44. **B**
6. **D**	19. **C**	32. **D**	45. **D**
7. **B**	20. **E**	33. **D**	46. **A**
8. **E**	21. **B**	34. **C**	47. **C**
9. **A**	22. **C**	35. **B**	48. **B**
10. **D**	23. **A**	36. **B**	49. **B**
11. **D**	24. **B**	37. **A**	50. **C**
12. **C**	25. **B**	38. **B**	
13. **B**	26. **E**	39. **C**	

ANSWERS EXPLAINED

The following explanations are keyed to the review portions of this book. The number in brackets after each explanation indicates the appropriate section in the Review of Major Topics (Part 2). If a problem can be solved using algebraic techniques alone, [algebra] appears after the explanation, and no reference is given for that problem in the Self-Evaluation Chart at the end of the test.

> In these solutions the following notation is used:
>
> i: calculator unnecessary
> a: calculator helpful or necessary
> g: graphing calculator helpful or necessary

1. g **(E)** Graph the function f in the standard window and observe the vertical asymptote at $x = -2$.

   An alternative solution is to factor the denominator of f as $(x + 2)(x - 2)$; cancel the factor $x - 2$ in the numerator and denominator so that $f(x) = \dfrac{1}{x+2}$; and recall that a function has a vertical asymptote when the denominator is zero and the numerator isn't. There is a hole in the graph of f at $x = 2$. [1.5]

2. i **(B)** The only pairs of parallel edges are the opposite sides of the square base. [2.2]

3. i **(E)** $\text{Log}(a^2 - b^2) = \log(a + b)(a - b) = \log(a + b) + \log(a - b)$. [1.4]

4. a **(D)** The sum of the roots is $\sqrt{2} + \sqrt{2} + (-\sqrt{3}) + \sqrt{5} \approx 1.414 + 1.414 - 1.732 + 2.236 \approx 3.3$. [1.2]

5. i **(A)** The slope of the first line is $-\dfrac{1}{2}$, and the slope of the second is $-\dfrac{a}{3}$. To be perpendicular, $-\dfrac{1}{2} = \dfrac{3}{a}$, or $a = -6$. [1.2]

6. g **(D)** Plot the graph of $6\sin x \cos x$ using ZTrig and observe that the max of this function is 3. (None of the other answer choices is close. If one were, you could use CALC/max to find the maximum value of the function.)

   An alternative solution is to recall that $2x = 2\sin x \cos x$, so that $6\sin x \cos x = 3\sin 2x$. Since the amplitude of $\sin 2x$ is 1, the amplitude of $3\sin 2x$ is 3. [1.3]

7. a **(B)** Put your calculator in radian mode. $f(2,3) = 2 \cdot \cos 3 \approx 2(-0.98999) \approx -1.98$. [2.1]

8. i **(E)** Since 5 and -1 are zeros, $x - 5$ and $x + 1$ are factors of $P(x)$, so their product $x^2 - 4x - 5$ is too. [1.2]

9. g **(A)** Enter the expression into your graphing calculator. [3.2]

10. g **(D)** Plot the graph of $y = \sin 2x$ in degree mode in an $x\varepsilon[10° \times 350°]$, $y\varepsilon[-2,2]$ window and observe that the graph crosses the axis 3 times.

An alternative explanation uses the fact that the function $\sin 2x$ has period $\dfrac{2\pi}{2} = \pi$ and the fact that $\sin 2x = 0$ when $2x = 0°, 180°, 360°, 540°, 720°, \ldots$, or when $x = 0°, 90°, 180°, 270°, 360°, \ldots$ Three values of x lie between $10°$ and $350°$. [1.3]

11. i (D) The third and seventh terms are 4 terms apart, and the difference between them is 8. Therefore, the common difference is $8 \div 4 = 2$. Since the first term is two terms prior to the third term, its value is $15 - 2(2) = 11$. [3.4]

12. i (C) Surface area $= 4\pi r^2$. Volume $= \dfrac{4}{3}\pi r^3$. Solve $4\pi^2 = \dfrac{4}{3}\pi r^3$ to get $r = 3$. [2.2]

13. a (B) Use the double angle formula to get $\sin 2x = 2 \sin x \cos x$. Since $\sin x = -\dfrac{3}{5}$,

$$\cos x = \pm\sqrt{1 - \sin^2 x} = \pm\sqrt{\dfrac{16}{25}} = \pm\dfrac{4}{5}.$$ The problem says that $\dfrac{3\pi}{2} < x < 2\pi$, so $\cos x = \dfrac{4}{5}$.

Substitute these values into the formula to get $\sin 2x = 2\left(-\dfrac{3}{5}\right)\left(\dfrac{4}{5}\right) = -\dfrac{24}{25}$. [1.3]

<u>Alternative Solution:</u> Plot the graph of $Y_1 = \sin x$ and $Y_2 = -0.6$ in a $x\varepsilon\left[\dfrac{3\pi}{2}, 2\pi\right]$, $y\varepsilon[-2.2]$ window, and find the point of intersection $(5.4, -0.6)$. Return to the home screen and enter $\sin(2x)$ to get -0.96. Change this to the fraction $-\dfrac{24}{25}$.

14. i (D) $s = r\theta$. $12 = r$. [1.3]

15. a (A) The domain consists of all numbers that make $3x^3 - 7 \geq 0$. Therefore,

$$x^3 \geq \dfrac{7}{3} \text{ and } x \geq 1.33. \text{ [1.1]}$$

16. i (C) Cofunctions of complementary angles are equal. Since x and y are complementary, tan and cot are cofunctions.

An alternative solution is to choose any values of x and y such that their sum is $90°$. (For example, $x = 40°$ and $y = 50°$.) Test the answer choices with your calculator in degree mode to see that only Choice C is true. [1.3]

17. g (B) Plot the graph of $y = x^3 + 5x + 1$ in the standard window and zoom in a couple of times to see that it crosses the x-axis only once. To make sure you are not missing anything, you should also plot the equation in a $x\varepsilon[-1,1]$, $y\varepsilon[-1,1]$ and an $x\varepsilon[-100,100]$, $y\varepsilon[-100,100]$ window.

An alternative solution is to use Descartes Rule of Signs, which indicates that the graph does not intersect the positive x-axis, but it does intersect the negative x-axis once. [1.2]

18. a (E) Complete the square in x and y: $(x^2 + 2x + 1) + (y^2 - 4y + 4) + z^2 = 10 + 1 + 4$, so $(x + 1)^2 + (y - 2)^2 + z^2 = 15$. Therefore, $r = \sqrt{15} \approx 3.87$. [2.2]

19. i **(C)** If you substitute 0 for x you get c, so c is the y-intercept of the graph. The graph and window indicate that this value is about 80. [1.2]

20. g **(E)** Plot the graph of $y = x(x - 3)(x + 2)$ in the standard window, and observe that the graph is above the x-axis when $-2 < x < 0$ or when $x > 3$.

An alternative solution is to find the zeros of the function $x(x - 3)(x + 2)$ as $x = 0, 3, -2$ and test points in the intervals established by these zeros. Points between -2 and 0 and greater than 3 satisfy the inequality. [1.2]

21. g **(B)** Use the program DPL to calculate the distance between the point $(0,0)$ and the line $3x - 4y = 10$ as 2. Therefore, the radius of the circle is 2, and its equation is $x^2 + y^2 = 4$. [2.2]

22. i **(C)**

$$\frac{f(x+t) - f(x)}{t} = \frac{3 - 2(x+t) + (x+t)^2 - (3 - 2x + x^2)}{t}$$

$$= \frac{3 - 2x - 2t + x^2 + 2xt + t^2 - 3 + 2x - x^2}{t}$$

$$= 2x - 2 + t. \quad [1.1]$$

> **CAUTION:** For calculus students only: This difference quotient looks like the definition of the derivative. However, no limit is taken, so don't jump at $f'(x)$, which is Choice D.

23. g **(A)** Enter x^3 into Y_1 and $x^2 + 1$ into Y_2. Then enter Y_1Y_2 into Y_3; $Y_1(Y_2)$ into Y_4; and $Y_2(Y_1)$ into Y_5. De-select Y_1 and Y_2. Inspection of Y_3, Y_4, and Y_5 shows that only Y_3 is symmetric about the origin.

An alternative solution is to define each of the three functions as $h(x)$ and check each against the definition of an odd function, $h(-x) = -h(x)$:

$$h(-x) = (-x)^3((-x)^2 + 1) = -x^5 - x^3 = -(x^5 + x^3)$$
$$= -h(x)$$
$$h(-x) = f((-x)^2 + 1) = (x^2 + 1)^3 \neq -h(x)$$
$$h(-x) = g((-x)^3) = ((-x)^3)^2 + 1 = x^6 + 1 \neq -h(x)$$

[1.1]

24. g **(B)** Since one particular man must be on the committee, the problem becomes: "Form a committee of 3 from 8 men." Calculate $8_nC_r3 = 56$. [3.1]

25. g **(B)** Graph MATH/NUM/max $(\sin(x), \cos(x))$ in the window $x \varepsilon [0, 2\pi]$ and $y \varepsilon [-2, 2]$. This graphs the exact function given, and the correct answer B will be evident.

An alternative solution is to sketch graphs of $\sin x$ and $\cos x$ on the same set of axes and highlight the graph of the maximum. [1.3]

26. a **(E)** Mean $= \dfrac{1 + 2 + 3 + 1 + 2 + 5 + x}{7} = \dfrac{14 + x}{7} = 3.\overline{27}$. Therefore, $x = 8.9$. [4.1]

27. a **(E)** Law of sines: $\dfrac{\frac{1}{3}}{\sqrt{5}} = \dfrac{\frac{3}{5}}{KL}$; $\dfrac{1}{3}KL = \dfrac{3\sqrt{5}}{5}$.

Therefore, $KL = \dfrac{9\sqrt{5}}{5} \approx 4.0$. [1.3]

28. i **(B)** Multiplication of matrices is possible only when the number of columns in the matrix on the left equals the number of rows in the matrix on the right. In that case, the product is a matrix with the number of rows in the left-hand matrix and the number of columns of the right-hand matrix. [3.3]

29. i **(C)** Relative to the statement, answer choice A is the converse, B is the inverse, C is the contrapositive, and D is another form of the inverse. Of these, the contrapositive is the logical equivalent of the original statement. [logic]

30. i **(D)** $(g \circ f)(x) = g(f(x)) = \sqrt{x-7}$. Therefore, $x \geq 7$. [1.1]

31. a **(C)** Law of cosines: $c^2 = 16 + 1 - 8 \cdot \dfrac{\sqrt{3}}{2} = 17 - 4\sqrt{3} \approx 10.07$. Therefore, $c \approx 3.2$. [1.3]

32. g **(D)** Use the standard window to graph $y < -\dfrac{3}{4}x$, by moving the cursor all the way left (past Y =) and keying Enter until a "lower triangle" is observed. The shaded portion of the graph will lie in all but Quadrant I.

An alternative solution is to graph the related equation $y = -\dfrac{3}{4}x$ and test points to determine which side of the line contains solutions to the inequality. This will indicate the quadrant that the graph does not enter. [1.2]

33. i **(D)** Fold the graph about the line $y = x$, and the resulting graph will be Choice D. [1.1]

34. a **(C)** Since these points are on a line, all slopes must be equal.

Slope $= \dfrac{-2-11}{5-(-2)} = \dfrac{-13}{7} = \dfrac{4.3-11}{x-(-2)}$. Cross-multiplying the far right equation gives

$-13(x+2) = 7(-6.7)$, and solving for x yields $x \approx 1.6$. [1.2]

35. i **(B)** Since the median is the middle value, it does not change if the number (quantity) of values above and below it remain the same. [4.1]

36. g **(B)** Plot the graph of $y = x^{-2/3}$ in the standard window and observe that the entire graph lies above the x-axis.

An alternative solution uses the fact that $x^{-2/3} = \dfrac{1}{x^{2/3}}$, and $x^{2/3} = (x^{1/3})^2$, so that for all values of x (except zero) y is positive. Therefore, the range of $y = x^{-2/3}$ is $y > 0$. [1.1]

37. a **(A)** Since $\dfrac{3\pi}{2} \le x \le 2\pi$, x is in the fourth quadrant. Taking $\cos^{-1}\dfrac{4}{5}$ on the calculator

yields first quadrant angle. Therefore, the fourth quadrant angle is $-\cos^{-1}\dfrac{4}{5}$. Enter

$\tan\left(-2\cos^{-1}\dfrac{4}{5}\right)$ and change to fractional form to get $-\dfrac{24}{7}$.

An alternative solution can be obtained from the fact that if $\cos x = \dfrac{4}{5}$ in Quadrant IV,

then $\tan x = -\dfrac{3}{4}$. Then use the double-angle formula for the tangent,

$\tan 2x = \dfrac{2\tan x}{1-\tan^2 x}$, and simplify: $\tan 2x = \dfrac{-3/2}{1-9/16} = -\dfrac{24}{7}$. [1.3]

38. i **(B)** There are 8 elements in the sample space of a coin being flipped 3 times. Of these elements, 7 contain at least 1 head and 3 (HHT, HTH, THH) contain 2 heads.

Probability = $\dfrac{3}{7}$. [4.2]

39. a **(C)** To find a unit vector parallel to $(2,-3,6)$, divide each component by

$\sqrt{2^2 + (-3)^2 + 6^2} = \pm 7$. There are two unit vectors, pointing in opposite directions,

that meet this requirement: $(0.29,-0.43,0.86)$ and $(-0.29,0.43,-0.86)$. Only the second one is an answer choice. [3.5]

40. i **(E)** To find the horizontal asymptote, you need to see what happens when x gets much larger ($\rightarrow\infty$) or much smaller ($\rightarrow-\infty$). With the numerator and denominator expanded,

$f(x) = \dfrac{2x^2 + 5x - 3}{x^2 + 6x + 9}$. Divide both by x^2 to see that y approaches 2 as x gets much bigger

or smaller. [1.5]

41. i **(B)** $f^2(x) = f(f(x)) = f\left(\dfrac{x}{x-1}\right) = \dfrac{\dfrac{x}{x-1}}{\dfrac{x}{x-1}-1} = x$

$f^3(x) = f(f^2(x)) = f(x) = \dfrac{x}{x-1}$. [3.4]

42. g (B) There are $\binom{15}{5}$ ways of selecting a committee of 5 out of 15 people (men and women). There are $\binom{6}{3}$ ways of selecting 3 men of 6, and for each of these, there are $\binom{9}{2}$ ways of selecting 2 women of 9. Therefore, there are $\binom{6}{3}\binom{9}{2}$ ways of selecting 3 men and 2 women. The probability of selecting 3 men and 2 women is

$$\frac{\binom{6}{3}\binom{9}{2}}{\binom{15}{5}} = \frac{240}{1001}.$$ Using the calculator command $nCr = \binom{n}{r}$, enter this expression and change to a fraction. [3.1, 4.2]

43. a (A) $2x + \sqrt{3} = (x + \sqrt{2}) + d$, and $5x - \sqrt{5} = (2x + \sqrt{3}) + d$. Eliminate d, and

$x + \sqrt{3} - \sqrt{2} = 3x - \sqrt{5} - \sqrt{3}$. Thus, $x = \dfrac{2\sqrt{3} - \sqrt{2} + \sqrt{5}}{2} \approx 2.14$. [3.4]

44. i (B) Substituting for y gives $x\left(\dfrac{4t}{t-3}\right) - 4x - 2\left(\dfrac{4t}{t-3}\right) - 4 = 0$, which simplifies to $12x - 12t + 12 = 0$. Then $x = t - 1$. [1.6]

45. i (D) The graph of the function is a parabola, and the equation of its axis of symmetry is $x = -\dfrac{b}{2a}$. Since the graph of $y = f(x - 3)$ is the translation of $y = f(x)$ 3 units to the right, $f(x - 3)$ will be symmetric about the y-axis when $f(x)$ is symmetric about -3. Therefore $-\dfrac{b}{2a} = -3$, or $b = 6a$. [1.2]

46. g (A) Use the change of base formula and graph $y = \dfrac{\log x}{\log 5}$ as Y_1. Graph $y = \ln(0.5x)$ as Y_2. The answer choices suggest a window of $x\varepsilon[0,10]$ and $y\varepsilon[0,2]$ Use the CALC/intersect to find the correct answer choice A.

An alternative solution converts the equations to exponential form: $x = 5^y$ and $\dfrac{1}{2}x = e^y$, or $5^y = 2e^y$. Taking the natural log of both sides gives $y\ln 5 = \ln 2 + y$, and solving for y yields $y = \dfrac{\ln 2}{\ln 5 - 1} \approx 1.14$. Finally substituting back to find x, $5^{1.14} \approx 6.24$. [1.4]

47. a (C) With your calculator in radian mode, plot the graphs of $Y_1 = \sin x$ and $Y_2 = 5 \cos x$ is a $x\varepsilon\left[\pi, \dfrac{3\pi}{2}\right]$ and $y\varepsilon[-5,5]$ window. The solution is the x-coordinate of the point of intersection. An alternative solution is to divide both sides of the equation by $\cos x$ to get $\tan x = 5$. Use your calculator to get $\tan^{-1} 5 = 1.373$. This is the first-quadrant solution, so add π. [1.3]

48. g (B) With your calculator in polar mode, graph $r = \dfrac{1}{\sin\theta + \cos\theta}$ in a $x\varepsilon[-3,3]$ and $y\varepsilon[-3,3]$ window. Let θ run from 0 to 2π in increments of 0.1. The graph of the function and the two axes form an isosceles right triangle of side length 1, so its area is 0.5.

An alternative solution is to multiply both sides of the equation by $\sin\theta + \cos\theta$ to get $r\sin\theta + r\cos\theta = 1$. Since $x = r\cos\theta$ and $y = r\sin\theta$, the rectangular form of the equation is $x + y = 1$. This line makes an isosceles right triangle of unit leg lengths with the axes, the area of which is $\dfrac{1}{2}(1)(1) = 0.5$ [2.1]

49. a (B) The diagonal of the base is $\sqrt{6^2 + 4^2} = \sqrt{52}$. The diagonal of the box is the hypotenuse of a right triangle with one leg $\sqrt{52}$ and the other leg 5. Let θ be the angle formed by the diagonal of the base and the diagonal of the box. $\tan\theta = \dfrac{5}{\sqrt{52}} \approx 0.69337$, and so $\theta = \tan^{-1} 0.69337 \approx 35°$. [1.3]

50. g (C) Plot $y = (2x + 1)^2$ in a window with $x\varepsilon[-2,2]$ and $t\varepsilon[-1,5]$, and observe that choice C is the only possible choice.

An alternative solution is to note that the graph of $y = (2x+1)^2 = 4\left(x + \dfrac{1}{2}\right)^2$ is a parabola with vertex at $\left(-\dfrac{1}{2}, 0\right)$ that opens up. The only answer choice with these properties is C. [1.2]

Test 2—Self-Evaluation Chart

Self-Evaluation Chart for Model Test 2

Subject Area	Questions and Review Section						Right	Number Wrong	Omitted
Algebra and Functions (21 questions)	1	3	4 ᵂ	5	8	15	5	1	0
	1.5	1.4	1.2	1.2	1.2	1.1			
	17	19	20	22	23 ᵂ	30	5	1	0
	1.2	1.2	1.2	1.1	1.1	1.1			
	32	33	34	36	40	44	5	0	0
	1.2	1.1	1.2	1.1	1.5	1.6			
	45 ᵂ	46	50				2	1	0
	1.2	1.4	1.2						
Trigonometry (11 questions)	6	10	13	14	16	25 ᵂ	5	1	0
	1.3	1.3	1.3	1.3	1.3	1.3			
	27	31	37	47	49 ᵂ		4	1	0
	1.3	1.3	1.3	1.3	1.3				
Coordinate and Three-Dimensional Geometry (6 questions)	2	7	12	18 ᵒ	21 ᵒ	48	4	0	2
	2.2	2.1	2.2	2.2	2.2	2.1			
Numbers and Operations (8 questions)	9	11	24	28 ᵂ	39 ᵒ	41	4	1	1
	3.2	3.4	3.1	3.3	3.5	3.4			
	42	43					2	0	0
	3.1	3.4							
Data Analysis, Statistics, and Probability (4 questions)	26	35	38 ᵂ	42	29 ᵂ		2	2	0
	4.1	4.1	4.2	4.2	Logic				
TOTALS							38	8	3

Evaluate Your Performance Model Test 2	
Rating	**Number Right**
Excellent	41–50
Very good	33–40
Above average	25–32
Average	15–24
Below average	Below 15

Calculating Your Score

Raw score R = number right $- \frac{1}{4}$ (number wrong), rounded = _____36_____

Approximate scaled score $S = 800 - 10(44 - R)$ = _____720_____

If $R \geq 44$, $S = 800$.

Answer Sheet
MODEL TEST 3

1 (A) (B) (C) (D)   14 (A) (B) (C) (D)   27 (A) (B) (C) (D)   40 (A) (B) (C) (D)
2 (A) (B) (C) (D)   15 (A) (B) (C) (D)   28 (A) (B) (C) (D)   41 (A) (B) (C) (D)
3 (A) (B) (C) (D)   16 (A) (B) (C) (D)   29 (A) (B) (C) (D)   42 (A) (B) (C) (D)
4 (A) (B) (C) (D)   17 (A) (B) (C) (D)   30 (A) (B) (C) (D)   43 (A) (B) (C) (D)
5 (A) (B) (C) (D)   18 (A) (B) (C) (D)   31 (A) (B) (C) (D)   44 (A) (B) (C) (D)
6 (A) (B) (C) (D)   19 (A) (B) (C) (D)   32 (A) (B) (C) (D)   45 (A) (B) (C) (D)
7 (A) (B) (C) (D)   20 (A) (B) (C) (D)   33 (A) (B) (C) (D)   46 (A) (B) (C) (D)
8 (A) (B) (C) (D)   21 (A) (B) (C) (D)   34 (A) (B) (C) (D)   47 (A) (B) (C) (D)
9 (A) (B) (C) (D)   22 (A) (B) (C) (D)   35 (A) (B) (C) (D)   48 (A) (B) (C) (D)
10 (A) (B) (C) (D)   23 (A) (B) (C) (D)   36 (A) (B) (C) (D)   49 (A) (B) (C) (D)
11 (A) (B) (C) (D)   24 (A) (B) (C) (D)   37 (A) (B) (C) (D)   50 (A) (B) (C) (D)
12 (A) (B) (C) (D)   25 (A) (B) (C) (D)   38 (A) (B) (C) (D)
13 (A) (B) (C) (D)   26 (A) (B) (C) (D)   39 (A) (B) (C) (D)

Model Test 3

ear out the preceding answer sheet. Decide which is the best choice by rounding your answer when appropriate. Blacken the corresponding space on the answer sheet. When finished, check your answers with those at the end of the test. For questions that you got wrong, note the sections containing the material that you must review. Also if you do not fully understand how you arrived at some of the correct answers, you should review the appropriate sections. Finally, fill out the self-evaluation chart on page 278 in order to pinpoint the topics that give you the most difficulty.

50 questions: 1 hour

Directions: Decide which answer choice is best. If the exact numerical value is not one of the answer choices, select the closest approximation. Fill in the oval on the answer sheet that corresponds to your choice.

Notes:
(1) You will need to use a scientific or graphing calculator to answer some of the questions.
(2) You will have to decide whether to put your calculator in degree or radian mode for some problems.
(3) All figures that accompany problems are plane figures unless otherwise stated. Figures are drawn as accurately as possible to provide useful information for solving the problem, except when it is stated in a particular problem that the figure is not drawn to scale.
(4) Unless otherwise indicated, the domain of a function is the set of all real numbers for which the functional value is also a real number.

Reference Information. The following formulas are provided for your information.

Volume of a right circular cone with radius r and height h: $V = \frac{1}{3}\pi r^2 h$

Lateral area of a right circular cone if the base has circumference c and slant height is l:

$S = \frac{1}{2}cl$

Volume of a sphere of radius r: $V = \frac{4}{3}\pi r^3$

Surface area of a sphere of radius r: $S = 4\pi r^2$

Volume of a pyramid of base area B and height h: $V = \frac{1}{3}Bh$

1. The slope of a line that is perpendicular to $3x + 2y = 7$ is

   (A) –2

   (B) $-\dfrac{3}{2}$

   (C) $\dfrac{2}{3}$

   (D) $\dfrac{3}{2}$

   (E) 2

   [handwritten: $2y = -3x+7$, $y = -\dfrac{3}{2}x$]

2. What is the remainder when $3x^4 - 2x^3 - 20x^2 - 12$ is divided by $x + 2$?

   (A) –60
   (B) –36
   (C) –28
   (D) –6
   (E) –4

3. If $1 - \dfrac{1}{x} = 2 - \dfrac{2}{x}$, then $3 - \dfrac{3}{x} =$

   (A) –3

   (B) $-\dfrac{1}{3}$

   (C) 0

   (D) $\dfrac{1}{3}$

   (E) 3

   [handwritten: $\left(1 - \dfrac{1}{x} = 2 - \dfrac{2}{x}\right)x$, $x - 1 = 2x - 2$, $1 = x$]

4. If $f(x) = 2 \ln x + 3$ and $g(x) = e^x$, then $f(g(3)) =$

   (A) 9
   (B) 11
   (C) 43.17
   (D) 47.13
   (E) 180.77

   [handwritten: $g(3) = e^3$, $f(e^3) = 2 \ln e^3 + 3$, $6 + 3$]

5. The domain of $f(x) = \log_{10}(\sin x)$ contains which of the following intervals?

   (A) $0 \le x \le \pi$

   (B) $-\dfrac{\pi}{2} \le x \le \dfrac{\pi}{2}$

   (C) $0 < x < \pi$

   (D) $-\dfrac{\pi}{2} < x < \dfrac{\pi}{2}$

   (E) $\dfrac{\pi}{2} < x < \dfrac{3\pi}{2}$

Model Test 3

6. Which of the following is the ratio of the surface area of the sphere with radius r to its volume?

   (A) $\dfrac{4}{r}$

   (B) $\dfrac{3}{r}$

   (C) $\dfrac{r}{4}$

   (D) $\dfrac{r}{\pi}$

   (E) $\dfrac{4}{\pi}$

$$\frac{4\pi r^2}{\frac{4}{3}\pi r^3}$$

$$\frac{4}{\frac{4}{3}r}$$

7. If the two solutions of $x^2 - 9x + c = 0$ are complex conjugates, which of the following describes all possible values of c?

   (A) $c = 0$

   (B) $c \neq 0$

   (C) $c < 9$

   (D) $c > \dfrac{81}{4}$

   (E) $c > 81$

$$b^2 - 4ac < 0$$
$$81 - 4(c) < 0$$
$$-4(c) < -81$$
$$c > \frac{81}{4}$$

8. If $\tan x = 3$, the numerical value of $\sqrt{\csc x}$ is

   (A) 0.32
   (B) 0.97
   (C) 1.03
   (D) 1.78
   (E) 3.16

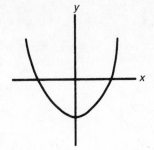

9. In the figure above, the graph of $y = f(x)$ has two transformations performed on it. First it is rotated 180° about the origin, and then it is reflected about the x-axis. Which of the following is the equation of the resulting curve?

   (A) $y = -f(x)$
   (B) $y = f(x + 2)$
   (C) $x = f(y)$
   (D) $y = f(x)$
   (E) none of the above

10. $\displaystyle \lim_{x \to \infty} \frac{3x^3 - 7x^2 + 2}{4x^2 - 3x - 1} =$

    (A) 0

    (B) $\dfrac{3}{4}$

    (C) 1

    (D) 3

    (E) ∞

11. As x increases from $-\dfrac{\pi}{4}$ to $\dfrac{3\pi}{4}$, the value of $\cos x$

    (A) always increases
    (B) always decreases
    (C) increases and then decreases
    (D) decreases and then increases
    (E) does none of the above

12. The vertical distance between the minimum and maximum values of the function $y = \left| -\sqrt{2} \sin \sqrt{3}x \right|$ is

    (A) 1.414
    (B) 1.732
    (C) 2.094
    (D) 2.828
    (E) 3.464

13. If the domain of $f(x) = -|x| + 2$ is $\{x: -1 \le x \le 3\}$, $f(x)$ has a minimum value when x equals

    (A) -1
    (B) 0
    (C) 1
    (D) 3
    (E) There is no minimum value.

14. What is the range of the function $f(x) = x^2 - 14x + 43$?

    (A) $x \le 7$
    (B) $x \ge 0$
    (C) $y \le -6$
    (D) $y \ge -6$
    (E) all real numbers

15. A positive rational root of the equation
$4x^3 - x^2 + 16x - 4 = 0$ is

(A) $\frac{1}{4}$

(B) $\frac{1}{2}$

(C) $\frac{3}{4}$

(D) 1

(E) 2

16. The norm of vector $\vec{V} = 3\vec{i} - \sqrt{2}\,\vec{j}$ is

(A) 4.24
(B) 3.61
(C) 3.32
(D) 2.45
(E) 1.59

17. If five coins are flipped and all the different ways they could fall are listed, how many elements of this list will contain more than three heads?

(A) 5
(B) 6
(C) 10
(D) 16
(E) 32

18. The seventh term of an arithmetic sequence is 5 and the twelfth term is -15. The first term of this sequence is

(A) 28
(B) 29
(C) 30
(D) 31
(E) 32

19. The graph of the curve represented by $\begin{cases} x = \sec\theta \\ y = \cos\theta \end{cases}$ is

(A) a line
(B) a hyperbola
(C) an ellipse
(D) a line segment
(E) a portion of a hyperbola

20. Point (3,2) lies on the graph of the inverse of $f(x) = 2x^3 + x + A$. The value of A is

(A) -54
(B) -15
(C) 15
(D) 18
(E) 54

21. If $f(x) = ax^2 + bx + c$ and $f(1) = 3$ and $f(-1) = 3$, then $a + c$ equals

(A) –3
(B) 0
(C) 2
(D) 3
(E) 6

$3 = a(1)^2 + b + c$
$3 = a + b + c$

$3 = a - b + c$

$a + b + c = a - b + c$
$b = -b$
$b = 0$

22. In $\triangle ABC$, $\angle B = 42°$, $\angle C = 30°$, and $AB = 100$. The length of BC is

(A) 47.6
(B) 66.9
(C) 133.8
(D) 190.2
(E) 193.7

$\dfrac{\sin 108}{X} = \dfrac{\sin 30}{100}$

23. If $4\sin x + 3 = 0$ on $0 \le x < 2\pi$, then $x =$

(A) –0.848
(B) 0.848
(C) 5.435
(D) 0.848 or 5.435
(E) 3.990 or 5.435

$\sin x = -\dfrac{3}{4}$
$\alpha = \sin^{-1}\dfrac{3}{4}$
$x = \pi + \alpha, \ 2\pi - \alpha$

24. The primary period of the function $f(x) = 2\cos^2 3x$ is

(A) $\dfrac{\pi}{3}$

(B) $\dfrac{\pi}{2}$

(C) π

(D) 2π

(E) 3π

25. In $a + bi$ form, the reciprocal of $2 + 6i$ is

(A) $\dfrac{1}{2} + \dfrac{1}{6}i$

(B) $-\dfrac{1}{16} + \dfrac{3}{16}i$

(C) $\dfrac{1}{16} + \dfrac{3}{16}i$

(D) $\dfrac{1}{20} - \dfrac{3}{20}i$

(E) $\dfrac{1}{20} + \dfrac{3}{20}i$

$\dfrac{1}{2+6i} \cdot \dfrac{(2-6i)}{(2-6i)}$

$\dfrac{2-6i}{4-36i^2} = \dfrac{2-6i}{4+36} = \dfrac{2-6i}{40}$

$= \dfrac{2(1-3i)}{40}$

$= \dfrac{1-3i}{20}$

26. A central angle of two concentric circles is $\dfrac{3\pi}{14}$.

The area of the large sector is twice the area of the small sector. What is the ratio of the lengths of the radii of the two circles?

(A) 0.25:1
(B) 0.50:1
(C) 0.67:1
(D) 0.71:1
(E) 1:1

27. If the region bounded by the lines $y = -\dfrac{4}{3}x + 4$,

$x = 0$, and $y = 0$ is rotated about the y-axis, the volume of the figure formed is

(A) 18.8
(B) 37.7
(C) 56.5
(D) 84.8
(E) 113.1

28. If there are known to be 4 broken transistors in a box of 12, and 3 transistors are drawn at random, what is the probability that none of the 3 is broken?

(A) 0.250
(B) 0.255
(C) 0.375
(D) 0.556
(E) 0.750

29. In order for the inverse of $f(x) = \sin 2x$ to be a function, the domain of f must be limited to

(A) $-\dfrac{\pi}{2} \le x \le \dfrac{\pi}{2}$

(B) $0 \le x \le \dfrac{\pi}{2}$

(C) $0 \le x \le \pi$

(D) $\dfrac{\pi}{4} \le x \le \dfrac{3\pi}{4}$

(E) $\dfrac{\pi}{2} \le x \le \pi$

30. Which of the following is a horizontal asymptote to

the function $f(x) = \dfrac{3x^4 - 7x^3 + 2x^2 + 1}{2x^4 - 4}$?

(A) $y = -3.5$
(B) $y = 0$
(C) $y = 0.25$
(D) $y = 0.75$
(E) $y = 1.5$

31. When a certain radioactive element decays, the amount at any time t can be calculated using the function $E(t) = ae^{\frac{-t}{500}}$, where a is the original amount and t is the elapsed time in years. How many years would it take for an initial amount of 250 milligrams of this element to decay to 100 milligrams?

(A) 125 years
(B) 200 years
(C) 458 years
(D) 496 years
(E) 552 years

32. If n is an integer, what is the remainder when $3x(2^{n+3}) - 4x(2^{n+2}) + 5x(2^{n+1}) - 8$ is divided by $x + 1$?

(A) −20
(B) −10
(C) −4
(D) 0
(E) The remainder cannot be determined.

33. Four men, A, B, C, and D, line up in a row. What is the probability that man A is at either end of the row?

(A) $\dfrac{1}{2}$
(B) $\dfrac{1}{3}$
(C) $\dfrac{1}{4}$
(D) $\dfrac{1}{6}$
(E) $\dfrac{1}{12}$

34. $\displaystyle\sum_{i=3}^{10} 5 =$

(A) 260
(B) 50
(C) 40
(D) 5
(E) none of these

35. The graph of $y^4 - 3x^2 + 7 = 0$ is symmetric with respect to which of the following?

 I. the *x*-axis
 II. the *y*-axis
 III. the origin

 (A) only I
 (B) only II
 (C) only III
 (D) only I and II
 (E) I, II, and III

36. In a group of 30 students, 20 take French, 15 take Spanish, and 5 take neither language. How many students take both French and Spanish?

 (A) 0
 (B) 5
 (C) 10
 (D) 15
 (E) 20

37. If $f(x) = x^2$, then $\dfrac{f(x+h) - f(x)}{h} =$

 (A) 0
 (B) *h*
 (C) 2*x*
 (D) 2*x* + *h*
 (E) $\dfrac{x^2}{h}$

38. The plane whose equation is $6x + 5y + 10z = 30$ forms a pyramid in the first octant with the coordinate planes. Its volume is

 (A) 15
 (B) 21
 (C) 30
 (D) 36
 (E) 45

39. Solve the equation $\sin 15x + \cos 15x = 0$. What is the sum of the three smallest positive solutions?

 (A) 0.16
 (B) 1.05
 (C) 1.10
 (D) 3.36
 (E) 16.49

40. Given the set of data 1, 1, 2, 2, 2, 3, 3, x, y, where x and y represent two different integers. If the mode is 2, which of the following statements must be true?

    (A) If $x = 1$ or 3, then y must = 2.
    (B) Both x and y must be > 3.
    (C) Either x or y must = 2.
    (D) It does not matter what values x and y have.
    (E) Either x or y must = 3, and the other must = 1.

USE THIS SPACE FOR SCRATCH WORK

41. If $f(x) = \sqrt{2x + 3}$ and $g(x) = x^2$, for what value(s) of x does $f(g(x)) = g(f(x))$?

    (A) –0.55
    (B) 0.46
    (C) 5.45
    (D) –0.55 and 5.45
    (E) 0.46 and 6.46

42. If $3x - x^2 \geq 2$ and $y^2 + y \leq 2$, then

    (A) $-1 \leq xy \leq 2$
    (B) $-2 \leq xy \leq 2$
    (C) $-4 \leq xy \leq 4$
    (D) $-4 \leq xy \leq 2$
    (E) $xy = 1, 2,$ or 4 only

43. In $\triangle ABC$, if $\sin A = \dfrac{1}{3}$ and $\sin B = \dfrac{1}{4}$, $\sin C =$

    (A) 0.14
    (B) 0.54
    (C) 0.56
    (D) 3.15
    (E) 2.51

44. The solution set of $\dfrac{|x-1|}{x} > 2$ is

    (A) $0 < x < \dfrac{1}{3}$
    (B) $x < \dfrac{1}{3}$
    (C) $x > \dfrac{1}{3}$
    (D) $\dfrac{1}{3} < x < 1$
    (E) $x > 0$

45. Suppose the graph of $f(x) = -x^3 + 2$ is translated 2 units right and 3 units down. If the result is the graph of $y = g(x)$, what is the value of $g(-1.2)$?

    (A) –33.77
    (B) –1.51
    (C) –0.49
    (D) 31.77
    (E) 37.77

$-(x-2)^3 + 2 - 3$

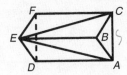

46. In the figure above, the bases, *ABC* and *DEF*, of the right prism are equilateral triangles of side *s*. The altitude of the prism *BE* is *h*. If a plane cuts the figure through points *A, C,* and *E*, two solids, *EABC*, and *EACFD*, are formed. What is the ratio of the volume of *EABC* to the volume of *EACFD*?

    (A) $\dfrac{1}{4}$

    (B) $\dfrac{1}{3}$

    (C) $\dfrac{\sqrt{3}}{4}$

    (D) $\dfrac{1}{2}$

    (E) $\dfrac{\sqrt{3}}{3}$

$\dfrac{1}{3} ABC \cdot BE$

47. A new machine can produce *x* widgets in *y* minutes, while an older one produces *u* widgets in *w* hours. If the two machines work together, how many widgets can they produce in *t* hours?

    (A) $t\left(\dfrac{x}{60y} + \dfrac{u}{w}\right)$

    (B) $t\left(\dfrac{60x}{y} + \dfrac{u}{w}\right)$

    (C) $60t\left(\dfrac{x}{y} + \dfrac{u}{w}\right)$

    (D) $t\left(\dfrac{y}{60x} + \dfrac{w}{u}\right)$

    (E) $t\left(\dfrac{x}{y} + \dfrac{60u}{w}\right)$

$X = y$

$\dfrac{X}{y}$

$60t\left(\dfrac{X}{y} + \dfrac{u}{60w}\right)$

$t\left(\dfrac{60X}{y} + \dfrac{u}{w}\right)$

$V = 60w$

$\dfrac{u}{60w}$

48. The length of the major axis of the ellipse
$3x^2 + 2y^2 - 6x + 8y - 1 = 0$ is

  (A) $\sqrt{3}$

  (B) $\sqrt{6}$

  (C) $2\sqrt{3}$

  (D) 4

  (E) $2\sqrt{6}$

49. If a coordinate system is devised so that the positive
y-axis makes an angle of 60° with the positive
x-axis, what is the distance between the points
with coordinates $(4,-3)$ and $(5,1)$?

  (A) 3.61
  (B) 3.87
  (C) 4.12
  (D) 4.58
  (E) 7.14

50. If $x - 7$ divides $x^3 - 3k^3x^2 - 13x - 7$, then $k =$

  (A) 1.19
  (B) 1.34
  (C) 1.72
  (D) 4.63
  (E) 5.04

STOP

If there is still time remaining, you may review your answers.

Answer Key
MODEL TEST 3

1. C	14. D	27. B	40. A
2. C	15. A	28. B	41. A
3. C	16. C	29. D	42. D
4. A	17. B	30. E	43. C
5. C	18. B	31. C	44. A
6. B	19. E	32. A	45. D
7. D	20. B	33. A	46. D
8. C	21. D	34. C	47. B
9. D	22. D	35. E	48. E
10. E	23. E	36. C	49. D
11. C	24. A	37. D	50. A
12. A	25. D	38. A	
13. D	26. D	39. C	

ANSWERS EXPLAINED

The following explanations are keyed to the review portions of this book. The number in brackets after each explanation indicates the appropriate section in the Review of Major Topics (Part 2). If a problem can be solved using algebraic techniques alone, [algebra] appears after the explanation, and no reference is given for that problem in the Self-Evaluation Chart at the end of the test.

In these solutions the following notation is used:

i: calculator unnecessary
a: calculator helpful or necessary
g: graphing calculator helpful or necessary

1. i **(C)** Transform the given equation into slope-intercept form: $y = -\dfrac{3}{2}x + \dfrac{7}{2}$ to see that

   the slope is $-\dfrac{3}{2}$. The slope of a perpendicular line is the negative reciprocal,

   or $\dfrac{2}{3}$. [1.2]

2. g **(C)** Let $f(x) = 3x^4 - 2x^3 - 20x^2 - 12$ and recall that $f(-2)$ is equal to the remainder upon division of $f(x)$ by $x + 2$. Enter $f(x)$ into Y_1, return to the Home Screen, and enter $Y_1(-2)$ to get the correct answer choice.

   An alternative solution is to use synthetic division to find the remainder.

   $$\begin{array}{r|rrrrr} -2 & 3 & -2 & -20 & 0 & -12 \\ & & -6 & 16 & 8 & -16 \\ \hline & 3 & -8 & -4 & 8 & -28 \end{array} = \text{remainder} \qquad [1.2]$$

3. i **(C)** Solve the equation by adding $\dfrac{2}{x} - 1$ to both sides and getting $\dfrac{2}{x} - \dfrac{1}{x} = 2 - 1$ or

   $\dfrac{1}{x} = 1$, so $x = 1$. Therefore, $3 - \dfrac{3}{x} = 3 - 3 = 0$. [algebra]

4. i **(A)** Since $g(3) = e^3$, $f(g(3)) = 2 \ln e^3 + 3$. $\ln e^3 = 3$. So $f(g(3)) = 6 + 3 = 9$. [1.4]

5. i **(C)** Since the domain of $\log_{10}$ is positive numbers, then the domain of f consists of values of x for which $\sin x$ is positive. This is only true for $0 < x < \pi$. [1.3,1.4]

6. i **(B)** $\dfrac{\text{Surface area of sphere}}{\text{Volume of sphere}} = \dfrac{4\pi r^2}{\frac{4}{3}\pi r^3} = \dfrac{12}{4r} = \dfrac{3}{r}$. [2.2]

7. g **(D)** Plot the graph of $y = x^2 - 9x$ in the standard window and observe that you must extend the window in the negative y direction to capture the vertex of the parabola. Since this vertex must lie above the x-axis for the solutions to be complex conjugates,

   c must be bigger than |minimum| = 20.25 = $\dfrac{81}{4}$. (The minimum is found using CALC/minimum.)

   An alternative solution is to use the fact that for the solutions to be complex conjugates, the discriminant $b^2 - 4ac = 81 - 4c < 0$, or $c > \dfrac{81}{4}$. [1.2]

8. g (C) Since $\tan x = 3$, x could be in the first or third quadrants. Since, however, $\sqrt{\csc x}$ is only defined when $\csc x \geq 0$, we need only consider x in the first quadrant. Thus,

we can enter $\sqrt{\left(\dfrac{1}{(\sin(\tan^{-1} 3))}\right)}$ to get the correct answer choice B. [1.3]

9. i (D) The two transformations put the graph right back where it started. [2.1]

10. g (E) Enter $(3x^3 - 7x^{2+2})/(4x^2 - 3x - 1)$ into Y_1. Enter TBLSET and set TblStart = 110 and ΔTbl = 10. Then enter TABLE and scroll down to larger and larger x values until you are convinced that Y_1 grows without bound.

An alternative solution is to divide the numerator and denominator by x^3 and then let $x \to \infty$. The numerator approaches 3 while the denominator approaches 0, so the whole fraction grows without bound. [1.2]

11. g (D) Plot the graph of $y = \cos x$ in an $x\varepsilon\left[-\dfrac{\pi}{4}, \dfrac{3\pi}{4}\right]$ and $y\varepsilon[-1,1]$ window and inspect

the graph to arrive at the correct answer choice C.

An alternative solution is to sketch the unit circle and observe that the x-coordinate of

a point increases as it moves from $-\dfrac{\pi}{4}$ to 0 and then decreases as it moves from 0

to $\dfrac{3\pi}{4}$. [1.3]

12. g (A) Plot the graph of $y = \left|-\sqrt{2}\sin\sqrt{3x}\right|$ using Ztrig. The minimum value of the function is clearly zero, and you can use CALC/maximum to establish 1.414 as the maximum value.

This function inside the absolute value is sinusoidal with amplitude $\sqrt{2} \approx 1.414$. The absolute value eliminates the bottom portion of the sinusoid, so this is the vertical distance between the maximum and the minimum as well. [1.3]

13. g (D) Plot the graph of $y = -|x| + 2$ on an $x\varepsilon[-1,3]$ and $y\varepsilon[-3,3]$ window. Examine the graph to see that its minimum value is achieved when $x = 3$.

An alternative solution is to realize that y is smallest when x is largest because of the negative absolute value. [1.6]

14. g (D) Plot the graph of $y = x^2 - 14x + 43$, and find its minimum value of –6. All values of y greater than or equal to –6 are in the range of f. [1.1]

15. g (A) Plot the graph of $y = 4x^3 - x^2 + 16x - 4$ in the standard window and zoom in once to get a clearer picture of the location of the zero. Use CALC/zero to determine that the zero is at $x = 0.25$.

An alternative solution is to use the Rational Roots Theorem to determine that the only

possible rational roots are $\pm 1, \pm 2, \pm 4, \pm \dfrac{1}{2}, \dfrac{1}{4}$. Synthetic division with these values in

turn eventually will yield the correct answer choice.

Another alternative solution is to observe that the left side of the equation can be factored: $4x^3 - x^2 + 16x - 4 = x^2(4x - 1) + 4(4x - 1) = (4x - 1)(x^2 + 1) = 0$. Since

$x^2 + 1$ can never equal zero, the only solution is $x = \dfrac{1}{4}$. [1.2]

Test 3—Answers Explained

16. a (C) $|\vec{V}| = \sqrt{3^2 + \left(-\sqrt{2}\right)^2} = \sqrt{9+2} = \sqrt{11} \approx 3.32.$ [3.5]

17. a (D) "More than 3" implies 4 or 5, and so the number of elements is $\binom{5}{4} + \binom{5}{5} = 6$. [3.1]

18. i (B) There are 5 common differences, d, between the seventh and twelfth terms, so $4d = -15 - 5 = -20$, and $d = -4$. The seventh term is the first term plus $6d$, so $5 = t_1 - 24$, and $t_1 = 29$. [3.4]

19. g (E) In parametric mode, plot the graph of $x = \sec t$ and $y = \cos t$ in the standard window to see that it looks like a portion of a hyperbola. You should verify that it is a portion of a hyperbola rather than the whole hyperbola by noting that $y = \cos t$ implies $-1 \leq y \leq 1$.

An alternative solution is to use the fact that secant and cosine are reciprocals so that elimination of the parameter t yields the equation $xy = 1$. This is the equation of a hyperbola and again, since $y = \cos t$ implies $-1 \leq y \leq 1$, the correct answer is E. [1.6]

20. i (B) If (3,2) lies on the inverse of f, (2,3) lies on f. Substituting in f gives $2 \cdot 2_3 + 2 + A = 3$. Therefore, $A = -15$. [1.1]

21. i (D) Substitute 1 for x to get $a + b + c = 3$. Substitute -1 for x to get $a - b + c = 3$. Add these two equations to get $a + b = 3$. [1.1]

22. a (D) $\angle A = 108°$. Law of sines: $\dfrac{a}{\sin 108°} = \dfrac{100}{\sin 30°}$. Therefore, $a = \dfrac{100 \sin 108°}{\sin 30°} = 190.2$. [1.3]

23. g (E) Plot the graph of $4 \sin x + 3$ in radian mode in an $x\varepsilon[0,2\pi]$ and $y\varepsilon[-2,8]$ window. Use CALC/zero twice to find the correct answer choice.

An alternative solution is to solve the equation for $\sin x$ and use your calculator, in radian mode: $\sin^{-1}\left(-\dfrac{3}{4}\right) = -0.848$.

Since this value is not between 0 and 2π, you must find the value of x in the required interval that has the same terminal side ($2\pi - 0.848 = 5.435$) as well as the third quadrant angle that has the same reference angle ($\pi + .848 = 3.990$). [1.3]

24. g (A) Plot the graph of $y = 2\cos(3x)^2$ using Ztrig, and trace one cycle to find its length of approximately 1.05. Then backsolve to get $\dfrac{\pi}{3}$.

An alternative solution uses the double angle formula for cosine: $\cos 2u = 2\cos^2 u - 1$, with $u = 3x$, to write $f(x) = \cos 6x + 1$. The primary period of this function is $\dfrac{2\pi}{6} = \dfrac{\pi}{3}$. [1.3]

25. a **(D)** Enter $\dfrac{1}{2+6i}$ into your calculator and press MATH/ENTER/ENTER/ to get FRACTIONAL real and imaginary parts. An alternative solution is to multiply the numerator and denominator of $\dfrac{1}{2+6i}$ by the complex conjugate $2-6i$:

$$\frac{1}{2+6i}\cdot\frac{2-6i}{2-6i}=\frac{2-6i}{4-(-36)}=\frac{1}{20}-\frac{3}{20}i \quad [3.2]$$

26. a **(D)** The measure $\dfrac{3\pi}{14}$ of the central angle is superfluous. Areas of similar figures are proportional to the squares of linear measures associated with those figures. Since the ratio of the areas is $1:2$, the ratio of the radii is $1:\sqrt{2}$, or approximately $0.71:1$. [1.3]

27. a **(B)** The line cuts the x-axis at 3 and the y-axis at 4 to form a right triangle that, when rotated about the y-axis, forms a cone with radius 3 and altitude 4.

$$\text{Volume} = \frac{1}{3}\pi r^2 h = \frac{1}{3}\pi(9)(4)=12\pi\approx 37.7. \quad [2.2]$$

28. a **(B)** Since there are 4 broken transistors, there must be 8 good ones. P(first pick is good) $= \dfrac{8}{12}$. Of the remaining 11 transistors, 7 are good, and so P(second pick is good) $= \dfrac{7}{11}$. Finally, P(third pick is good) $= \dfrac{6}{10}$. Therefore, P(all three are good) $=$

$$\frac{8}{12}\cdot\frac{7}{11}\cdot\frac{6}{10}=\frac{14}{55}\approx 0.255.$$

Alternative Solution: Note that there are $\binom{8}{3}=56$ ways to select 3 good transistors.

There are $\binom{12}{3}=220$ ways to select any 3 transistors. P(3 good ones) $=$

$$\frac{56}{220}=\frac{14}{55}\approx 0.255. \quad [4.2]$$

29. g **(D)** Plot the graph of $y=\sin 2x$ using Ztrig, and recall that for the inverse of a function to be a function, any horizontal line can only cross the graph of the function in one point. Using TRACE, you can see that this only occurs when $\dfrac{\pi}{4}\leq x\leq\dfrac{3\pi}{4}$.

An alternative solution is to use your knowledge of the sine graph and the fact that $\sin 2x$ has period $\dfrac{2\pi}{2}=\pi$. The function $\sin 2x$ increases from 0 to 1 on $\left[0,\dfrac{\pi}{4}\right]$, decreases from 1 to 0 on $\left[\dfrac{\pi}{4},\dfrac{\pi}{2}\right]$, from 0 to -1 on $\left[\dfrac{\pi}{2},\dfrac{3\pi}{4}\right]$, and finally from -1

back to 0 on $\left[\dfrac{3\pi}{4}, \pi\right]$. Each y value between 1 and -1 is achieved exactly once

between $\dfrac{\pi}{4}$ and $\dfrac{3\pi}{4}$. [1.1, 1.3]

30. i **(E)** Divide each term in the numerator and denominator by x^4. As x increases or decreases, all but the first terms in the numerator and denominator approach zero,

leaving a ratio of $\dfrac{3}{2}$. [1.5]

31. g **(C)** Plot the graphs of $Y_1 = 250e^{-t/500}$ and $Y_2 = 100$. Using the answer choices and information in the problem, set a $x\varepsilon[100,600]$ and $y\varepsilon[50,300]$ window, and find the point when Y_1 and Y_2 intersect. The solution is the x-coordinate of this point.

An alternative method is to substitute 250 for a and 100 for E to get $100 = 250e^{\frac{-t}{500}}$.

Divide both sides by 250. Take the ln of both sides of the equation. Finally, multiply

both sides by -500 to get $t = -500 \ln \dfrac{2}{5} \approx 458$. [1.4]

32. i **(A)** According to the Remainder Theorem, simply substitute -1 for x:
$3(-1) - 4(1) + 5(-1) - 8 = -20$. [1.2]

33. i **(A)** Since there are 4 positions man A is at one end in half of the arrangements . Therefore, p(man A is in an end

seat) $= \dfrac{1}{2}$. [4.2]

34. i **(C)** The summation indicates that 5 be summed 8 times (when $i = 3,4, \ldots, 11$), so the sum is $8 \times 5 = 40$. [3.4]

35. g **(E)** Graph $y = \sqrt[4]{3x^2 + 7}$ and $y = -\sqrt[4]{3x^2 + 7}$ in a standard window. Observe the graph is symmetrical with respect to the x-axis, the y-axis, and the origin.

An alternative solution is to observe that if x is replaced by $-x$; if y is replaced by $-y$, or if both replacements take place the equation is unaffected. Therefore, all three symmetries are present. [1.1]

36. i **(C)** From the Venn diagram below you get the following equations:

$$
\begin{array}{ll}
a + b + c + d & = 30 \qquad (1) \\
b + c & = 20 \qquad (2) \\
c + d & = 15 \qquad (3) \\
a & = 5
\end{array}
$$

Subtract equation (2) from equation (1): $a + d = 10$. Since $a = 5$, $d = 5$. Substituting 5 for d in equation (3) leaves $c = 10$. [3.1]

All students

37. i (D) First note that $f(x + h) = (x + h)^2 = x^2 + 2xh + h^2$. So $f(x + h) - f(x) = 2xh + h^2$. Divide this expression by h to get the correct answer choice. [1.1]

38. a (A) The plane cuts the *x*-axis at 5, the *y*-axis at 6, and the *z*-axis at 3. The base is a right triangle with area $\approx \frac{1}{2}(5)(6) = 15$. $V = \frac{1}{3}Bh = \frac{1}{3}(15)(3) = 15$. [2.2]

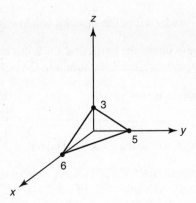

39. g (C) Graph $y = \sin 15x + \cos 15x$ in an $x\varepsilon\left[0, \frac{\pi}{5}\right]$ and $y\varepsilon[-2,2]$ window. Use CALC/zero

to find the far left zero, return to the Home Screen, and store it in A. Then use CALC/zero to find the middle zero, return to the Home Screen, and store it in B. Finally, use CALC/zero to find the far right zero, return to the Home Screen, and enter A + B + X to get 1.099. . . .

An alternative solution is to rewrite the equation as $\tan 15x = -1$. The three smallest

positive solutions for $15x$ are $\frac{3\pi}{4}, \frac{7\pi}{4}, \frac{11\pi}{4}$, so the three smallest positive solutions

for x are $\frac{3\pi}{60}, \frac{7\pi}{60}, \frac{11\pi}{60}$, and their sum is $\frac{7\pi}{20} \approx 1.10$. [1.3]

40. i (A) The number of 2s must exceed the number of other values. Some of the choices can be eliminated. B: one integer could be 2. C: *x* or *y* could be 2, but not necessarily. D and E: if $x = 3$ and $y = 1$, there will be no mode. Therefore, Choice A is the answer. [4.1]

41. g (A) Enter $\sqrt{2x + 3}$ into Y_1, x^2 into Y_2, $Y_1(Y_2(X))$ into Y_3, and $Y_2(Y_1(X))$ into Y_4. De-select Y_1 and Y_2, and graph Y_3 and Y_3 in an $x\varepsilon[-1.5,2]$ and $y\varepsilon[0,5]$ window (because $2x + 3 \geq 0$). Use CALC/intersect to find the correct answer choice A.

An alternative solution is to evaluate $f(g(x)) = \sqrt{2x^3 + 3}$ and $g(f(x)) = 2x + 3$, set the two equal, square both sides, and solve the resulting quadratic:
$2x^2 + 3 = (2x + 3)^2 = 4x^2 + 12x + 9$ or $x^2 + 6x + 3 = 0$. The Quadratic Formula yields

$x = \dfrac{-6 + \sqrt{24}}{2}$ or $x = \dfrac{-6 - \sqrt{24}}{2}$. However, the second is not in the domain of *g*, so

$x = \dfrac{-6 + \sqrt{24}}{2} \approx -0.55$ is the only solution. [1.1]

42. i **(D)** Solve the first inequality to get $1 \le x \le 2$. Solve the second inequality to get $-2 \le y \le 1$. The smallest product xy possible is -4, and the largest product xy possible is $+2$. [1.2]

43. a **(C)** Use your calculator to evaluate $\sin\left(180° - \sin^{-1}\frac{1}{3} - \sin^{-1}\frac{1}{4}\right) \approx 0.56$. [1.3]

44. g **(A)** Graph $y = \frac{|x-1|}{x}$ and $y = 2$ in the standard window. There is a vertical asymptote at $x = 0$, and the curve is above the horizontal $y = 2$ just to the right of 0. Answer choice A is the only possibility, but to be sure, use CALC/intersect to determine that the point of intersection is at $x = \frac{1}{3}$.

An alternative solution is to use the associated equation $\frac{|x-1|}{x} = 2$ to find boundary values and then test points. Since the left side is undefined when $x = 0$, zero is one boundary value. Multiply both sides by x to get $|x - 1| = 2x$ and analyze the two cases for absolute value. If $x - 1 \ge 0$, then $x - 1 = 2x$ so $x = -1$, which is impossible because $x \ge 1$. Therefore, $x - 1 < 0$, so $1 - x = 2x$ or $x = \frac{1}{3}$ is the other boundary value. Testing

points in the intervals $(-\infty, 0)$, $\left(0, \frac{1}{3}\right)$, and $\left(\frac{1}{3}, \infty\right)$ yields the correct answer choice A. [1.2]

45. g **(D)** Translating f 2 units right and 3 units down results in $g(x) = -(x - 2)^3 - 1$. Then $g(-1.2) \approx 31.77$. [2.1]

46. i **(D)** The volume of the prism is area of base times height. The figure *EABC* is a pyramid with base triangle *ABC* and height *BE*, the same as the base and height of the prism. The volume of the pyramid is $\frac{1}{3}$ (base times height), $\frac{1}{3}$ the volume of the prism. Therefore, the other solid, *EACFD*, is $\frac{2}{3}$ the volume of the prism. The ratio of the volumes is $\frac{1}{2}$. [2.2]

47. i **(B)** Since the new machine produces x widgets in y minutes, it can produce $\frac{60x}{y}$ widgets per hour. The old machine produces $\frac{u}{w}$ widgets per hour. Adding these and multiplying by t yields the correct answer choice B. [algebra]

48. i **(E)** Get the center axis form of the equation by completing the square:
$3x^2 - 6x + 2y^2 + 8y = 1$
$3(x^2 - 2x + 1) + 2(y^2 + 4y + 4) = 1 + 3 + 8 = 12$, which leads to
$\frac{3(x-1)^2}{12} + \frac{2(y+2)^2}{12} = 1$ and finally to $\frac{(x-1)^2}{4} + \frac{(y+2)^2}{6} = 1$.

Thus, half the major axis is $\sqrt{6}$, making the major axis $2\sqrt{6}$. [2.1]

49. a **(D)** From the figure it can be seen that the lines drawn parallel to the axes and the line through the two points form a triangle with one angle of 120° and adjacent sides of 4 and 1. From the law of cosines,

$$d_2 = 1 + 16 - 2(1)(4) \cos 120°$$
$$= 17 + 4$$

Thus, the distance between the two points $d = \sqrt{21} \approx 4.58$. [1.3]

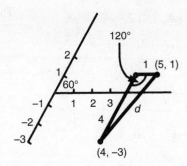

50. a **(A)** According to the factor theorem, substituting 7 for x yields 0. Therefore,

$$7^3 - 3(7)^2k^3 - 13(7) - 7 = 0 \text{, or } k = \sqrt[3]{\frac{245}{147}} \approx 1.19. \text{ [1.2]}$$

Self-Evaluation Chart for Model Test 3

Subject Area	Questions and Review Section						Right	Number Wrong	Omitted
Algebra and Functions (24 questions)	1 1.2	2 1.2	3	4 1.4	5 1.4	7 1.2	6	0	0
	10 1.2	13 1.6	14 1.1	15 1.2	19 1.6	20 1.1	6	0	0
	21 1.1	29 1.1	30 1.5	31 1.4	32 w 1.2	35 w 1.1	4	2	0
	37 1.1	41 1.1	42 o 1.2	44 w 1.2	47 o	50 o 1.2	2	1	3
Trigonometry (12 questions)	5 1.3	8 1.3	11 1.3	12 1.3	22 1.3	23 1.3	6	0	0
	24 1.3	26 w 1.3	29 w 1.3	39 w 1.3	43 1.3	49 o 1.3	2	3	1
Coordinate and Three-Dimensional Geometry (7 questions)	6 2.2	9 2.1	27 o 2.2	38 o 2.2	45 2.1	46 o 2.2	3	0	3
	48 o 2.1						0	0	1
Numbers and Operations (6 questions)	16 3.5	17 3.1	18 3.4	25 3.2	34 3.4	36 w 3.1	5	1	0
Data Analysis, Statistics, and Probability (3 questions)	28 4.2	33 4.2	40 4.1				3	0	0
TOTALS							37	7	8

Evaluate Your Performance Model Test 3	
Rating	**Number Right**
Excellent	41–50
Very good	33–40
Above average	25–32
Average	15–24
Below average	Below 15

Calculating Your Score

Raw score R = number right – $\frac{1}{4}$ (number wrong), rounded = _____35_____

Approximate scaled score $S = 800 - 10(44 - R)$ = _____710_____

If $R \geq 44$, $S = 800$.

Answer Sheet

MODEL TEST 4

1 Ⓐ Ⓑ Ⓒ Ⓓ	14 Ⓐ Ⓑ Ⓒ Ⓓ	27 Ⓐ Ⓑ Ⓒ Ⓓ	40 Ⓐ Ⓑ Ⓒ Ⓓ
2 Ⓐ Ⓑ Ⓒ Ⓓ	15 Ⓐ Ⓑ Ⓒ Ⓓ	28 Ⓐ Ⓑ Ⓒ Ⓓ	41 Ⓐ Ⓑ Ⓒ Ⓓ
3 Ⓐ Ⓑ Ⓒ Ⓓ	16 Ⓐ Ⓑ Ⓒ Ⓓ	29 Ⓐ Ⓑ Ⓒ Ⓓ	42 Ⓐ Ⓑ Ⓒ Ⓓ
4 Ⓐ Ⓑ Ⓒ Ⓓ	17 Ⓐ Ⓑ Ⓒ Ⓓ	30 Ⓐ Ⓑ Ⓒ Ⓓ	43 Ⓐ Ⓑ Ⓒ Ⓓ
5 Ⓐ Ⓑ Ⓒ Ⓓ	18 Ⓐ Ⓑ Ⓒ Ⓓ	31 Ⓐ Ⓑ Ⓒ Ⓓ	44 Ⓐ Ⓑ Ⓒ Ⓓ
6 Ⓐ Ⓑ Ⓒ Ⓓ	19 Ⓐ Ⓑ Ⓒ Ⓓ	32 Ⓐ Ⓑ Ⓒ Ⓓ	45 Ⓐ Ⓑ Ⓒ Ⓓ
7 Ⓐ Ⓑ Ⓒ Ⓓ	20 Ⓐ Ⓑ Ⓒ Ⓓ	33 Ⓐ Ⓑ Ⓒ Ⓓ	46 Ⓐ Ⓑ Ⓒ Ⓓ
8 Ⓐ Ⓑ Ⓒ Ⓓ	21 Ⓐ Ⓑ Ⓒ Ⓓ	34 Ⓐ Ⓑ Ⓒ Ⓓ	47 Ⓐ Ⓑ Ⓒ Ⓓ
9 Ⓐ Ⓑ Ⓒ Ⓓ	22 Ⓐ Ⓑ Ⓒ Ⓓ	35 Ⓐ Ⓑ Ⓒ Ⓓ	48 Ⓐ Ⓑ Ⓒ Ⓓ
10 Ⓐ Ⓑ Ⓒ Ⓓ	23 Ⓐ Ⓑ Ⓒ Ⓓ	36 Ⓐ Ⓑ Ⓒ Ⓓ	49 Ⓐ Ⓑ Ⓒ Ⓓ
11 Ⓐ Ⓑ Ⓒ Ⓓ	24 Ⓐ Ⓑ Ⓒ Ⓓ	37 Ⓐ Ⓑ Ⓒ Ⓓ	50 Ⓐ Ⓑ Ⓒ Ⓓ
12 Ⓐ Ⓑ Ⓒ Ⓓ	25 Ⓐ Ⓑ Ⓒ Ⓓ	38 Ⓐ Ⓑ Ⓒ Ⓓ	
13 Ⓐ Ⓑ Ⓒ Ⓓ	26 Ⓐ Ⓑ Ⓒ Ⓓ	39 Ⓐ Ⓑ Ⓒ Ⓓ	

Model Test 4

Tear out the preceding answer sheet. Decide which is the best choice by rounding your answer when appropriate. Blacken the corresponding space on the answer sheet. When finished, check your answers with those at the end of the test. For questions that you got wrong, note the sections containing the material that you must review. Also if you do not fully understand how you arrived at some of the correct answers, you should review the appropriate sections. Finally, fill out the self-evaluation chart on page 302 in order to pinpoint the topics that give you the most difficulty.

50 questions: 1 hour

Directions: Decide which answer choice is best. If the exact numerical value is not one of the answer choices, select the closest approximation. Fill in the oval on the answer sheet that corresponds to your choice.

Notes:
(1) You will need to use a scientific or graphing calculator to answer some of the questions.
(2) You will have to decide whether to put your calculator in degree or radian mode for some problems.
(3) All figures that accompany problems are plane figures unless otherwise stated. Figures are drawn as accurately as possible to provide useful information for solving the problem, except when it is stated in a particular problem that the figure is not drawn to scale.
(4) Unless otherwise indicated, the domain of a function is the set of all real numbers for which the functional value is also a real number.

Reference Information. The following formulas are provided for your information.

Volume of a right circular cone with radius r and height h: $V = \dfrac{1}{3}\pi r^2 h$

Lateral area of a right circular cone if the base has circumference c and slant height is l:

$S = \dfrac{1}{2}cl$

Volume of a sphere of radius r: $V = \dfrac{4}{3}\pi r^3$

Surface area of a sphere of radius r: $S = 4\pi r^2$

Volume of a pyramid of base area B and height h: $V = \dfrac{1}{3}Bh$

1. If point (a,b) lies on the graph of function f, which of the following points must lie on the graph of the inverse f?

   (A) (a,b)
   (B) $(-a,b)$
   (C) $(a,-b)$
   (D) (b,a)
   (E) $(-b,-a)$

2. Harry had grades of 70, 80, 85, and 80 on his quizzes. If all quizzes have the same weight, what grade must he get on his next quiz so that his average will be 80?

   (A) 85
   (B) 90
   (C) 95
   (D) 100
   (E) more than 100

   $$\frac{70+80+85+x+80}{5} = 80$$

3. Which of the following is an asymptote of

   $$f(x) = \frac{x^2+3x+2}{x+2} \cdot \tan \pi x \ ?$$

   (A) $x = -2$
   (B) $x = -1$
   (C) $x = \dfrac{1}{2}$
   (D) $x = 1$
   (E) $x = 2$

   $$\frac{(x+2)(x+1)}{x+2}$$

   $$(x+1)(\tan \pi x)$$

4. If $\log_b x = p$ and $\log_b y = q$, then $\log_b xy =$

   (A) pq
   (B) $p + q$
   (C) $\dfrac{p}{q}$
   (D) $p - q$
   (E) p^q

   $$\log_b x + \log_b y$$

5. The sum of the roots of $3x^3 + 4x^2 - 4x = 0$ is

   (A) $-\dfrac{4}{3}$
   (B) $-\dfrac{3}{4}$
   (C) 0
   (D) $\dfrac{4}{3}$
   (E) 4

   $$\frac{2}{3}$$

   $$-2$$

6. If $f(x) = x - \dfrac{1}{x}$, then $f(a) + f\left(\dfrac{1}{a}\right) =$

(A) 0

(B) $2a - \dfrac{2}{a}$

(C) $a - \dfrac{1}{a}$

(D) $\dfrac{a^4 - a^2 + 1}{a(a^2 - 1)}$

(E) 1

$$a - \frac{1}{a} + \left(\frac{\frac{1}{a}}{} - \frac{1}{\frac{1}{a}}\right)$$

$$a - \frac{1}{a} + \frac{1}{a} - a$$

7. If $f(x) = x^4 - 4x^3 + 6x^2 - 4x + 2$, then $f(2) - f\left(\sqrt{2}\right) =$

(A) 0.97

(B) 0.86

(C) 1.03

(D) 1.42

(E) 1.73

8. If $f(x) \geq 0$ for all x, then $f(2 - x)$ is

(A) ≥ -2

(B) ≥ 0

(C) ≥ 2

(D) ≤ 0

(E) ≤ 2

9. How many four-digit numbers can be formed from the numbers 0, 2, 4, 8 if no digit is repeated?

(A) 18

(B) 24

(C) 27

(D) 36

(E) 64

10. If $x - 1$ is a factor of $x^2 + ax - 4$, then a has the value

(A) 4

(B) 3

(C) 2

(D) 1

(E) none of the above

$$1^2 + a - 4 = 0$$

$$a - 3 = 0$$

11. If 10 coins are to be flipped and the first 5 all come up heads, what is the probability that exactly 3 more heads will be flipped?

(A) 0.0439

(B) 0.1172

(C) 0.1250

(D) 0.3125

(E) 0.6000

$5 C 3$

$5 C 5$ ⵒⵒⵒ HHHTT

120

12. If $i = \sqrt{-1}$ and n is a positive integer, which of the following statements is FALSE?

    (A) $i^{4n} = 1$
    (B) $i^{4n+1} = -i$
    (C) $i^{4n+2} = -1$
    (D) $i^{n+4} = i^n$
    (E) $i^{4n+3} = -i$

13. If $\log_r 3 = 7.1$, then $\log_r \sqrt{3} =$

    (A) 2.66
    (B) 3.55
    (C) $\dfrac{\sqrt{3}}{r}$

    (D) $\dfrac{7.1}{r}$

    (E) $\sqrt[r]{7.1}$

14. If $f(x) = 4x^2$ and $g(x) = f(\sin x) + f(\cos x)$, then $g(23°)$ is

    (A) 1
    (B) 4
    (C) 4.29
    (D) 5.37
    (E) 8

15. What is the sum of the roots of the equation
    $\left(x - \sqrt{2}\right)\left(x^2 - \sqrt{3}x + \pi\right) = 0$?

    (A) −0.315
    (B) −0.318
    (C) 1.414
    (D) 3.15
    (E) 4.56

16. Which of the following equations has (have) graphs consisting of two perpendicular lines?

    I. $xy = 0$
    II. $|y| = |x|$
    III. $|xy| = 1$

    (A) only I
    (B) only II
    (C) only III
    (D) only I and II
    (E) I, II, and III

17. A line, *m*, is parallel to a plane, *X*, and is 6 inches from *X*. The set of points that are 6 inches from *m* and 1 inch from *X* form

(A) a line parallel to *m*
(B) two lines parallel to *m*
(C) four lines parallel to *m*
(D) one point
(E) the empty set

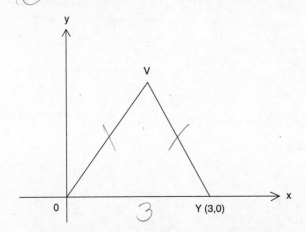

18. In the figure above, if *VO = VY*, what is the slope of segment *VO*?

(A) $-\sqrt{3}$

(B) $-\dfrac{\sqrt{3}}{2}$

(C) $\dfrac{\sqrt{3}}{2}$

(D) $\sqrt{3}$

(E) Cannot be determined from the given information.

19. A cylindrical bar of metal has a base radius of 2 and a height of 9. It is melted down and reformed into a cube. A side of the cube is

(A) 2.32
(B) 3.84
(C) 4.84
(D) 97.21
(E) 113.10

20. The graph of $y = (x + 2)(2x - 3)$ can be expressed as a set of parametric equations. If $x = 2t - 2$ and $y = f(t)$, then $f(t) =$

(A) $2t(4t - 5)$
(B) $(2t - 2)(4t - 7)$
(C) $2t(4t - 7)$
(D) $(2t - 2)(4t - 5)$
(E) $2t(4t + 1)$

USE THIS SPACE FOR SCRATCH WORK

$f(t) = (2t - 2 + 2)(2(2t - 2) - 3)$

$= 2t(4t - 4 - 3)$

$= 2t(4t - 7) =$

21. If points $\left(\sqrt{2}, y_1\right)$ and $\left(-\sqrt{2}, y_2\right)$ lie on the graph of $y = x^3 + ax^2 + bx + c$, and $y_1 - y_2 = 3$, then $b =$

(A) 1.473
(B) 1.061
(C) −0.354
(D) −0.939
(E) −2.167

$y_1 = (\sqrt{2})^3 + a(\sqrt{2})^2 + \sqrt{2}b + c$ $y_2 = -2\sqrt{2} + 4a - \sqrt{2}b + c$

$y_1 = 2\sqrt{2} + 4a + \sqrt{2}b + c$

$2\sqrt{2} + 4a + \sqrt{2}b + c - (-2\sqrt{2} + 4a - \sqrt{2}b + c) = 3$

$2\sqrt{2} + 2\sqrt{2} + \sqrt{2}b + \sqrt{2}b = 3$

$4\sqrt{2} + 2\sqrt{2}b = 3$

22. Rent-a-Rek has 27 cars available for rental. Twenty of these are compact, and 7 are midsize. If two cars are selected at random, what is the probability that both are compact?

(A) 0.0576
(B) 0.0598
(C) 0.481
(D) 0.521
(E) 0.541

$\dfrac{20}{27} \cdot \dfrac{19}{26}$

23. If a and b are real numbers, with $a > b$ and $|a| < |b|$, then

(A) $a > 0$
(B) $a < 0$
(C) $b > 0$
(D) $b < 0$
(E) none of the above

24. If $[x]$ is defined to represent the greatest integer less than or equal to x, and $f(x) = \left| x - [x] - \dfrac{1}{2} \right|$, the maximum value of $f(x)$ is

(A) −1
(B) 0
(C) $\dfrac{1}{2}$
(D) 1
(E) 2

25. $\lim\limits_{x \to 2} \dfrac{x^3 - 8}{x^2 - 4} =$

(A) 0
(B) 1
(C) 2
(D) 3
(E) ∞

26. A right circular cone whose base radius is 4 is inscribed in a sphere of radius 5. What is the ratio of the volume of the cone to the volume of the sphere?

(A) 0.222 : 1
(B) 0.256 : 1
(C) 0.288 : 1
(D) 0.333 : 1
(E) 0.864 : 1

h = 9

$V_{cone} = \frac{1}{3}\pi r^2 h$ $V_{sphere} = \frac{4}{3}\pi r^3$

$= 150.7$ $= 523$

27. If $x_0 = 1$ and $x_{n+1} = \sqrt[3]{2x_n}$, then $x_3 =$

(A) 1.260
(B) 1.361
(C) 1.396
(D) 1.408
(E) 1.412

$X_2 =$

28. The y-intercept of $y = \left| \sqrt{2} \csc 3\left(x + \dfrac{\pi}{5}\right) \right|$ is

(A) 0.22
(B) 0.67
(C) 1.41
(D) 1.49
(E) 4.58

29. If the center of the circle $x^2 + y^2 + ax + by + 2 = 0$ is point $(4, -8)$, then $a + b =$

(A) –8
(B) –4
(C) 4
(D) 8
(E) 24

$(x - 4)^2 + (y + 8)^2$

$x^2 + 16 - 8x \qquad y^2 + 64 + 16y$

30. If $p(x) = 3x^2 + 9x + 7$ and $p(a) = 2$, then $a =$

(A) only 0.736
(B) only –2.264
(C) 0.736 or 2.264
(D) 0.736 or –2.264
(E) –0.736 or –2.264

$2 = 3x^2 + 9x + 7$

$0 = 3x^2 + 9x + 5$

31. If i is a root of $x^4 + 2x^3 - 3x^2 + 2x - 4 = 0$, the product of the real roots is

(A) –4
(B) –2
(C) 0
(D) 2
(E) 4

32. If $\sin A = \dfrac{3}{5}, 90° \leq A \leq 180°, \cos B = \dfrac{1}{3}$, and

$270° \leq B \leq 360°$, $\sin(A + B) =$

(A) –0.832
(B) –0.554
(C) –0.333
(D) 0.733
(E) 0.954

33. If a family has three children, what is the probability that at least one is a boy?

(A) 0.875
(B) 0.67
(C) 0.5
(D) 0.375
(E) 0.25

34. If sec 1.4 = x, find the value of csc(2 Arctan x).

(A) 0.33
(B) 0.87
(C) 1.00
(D) 1.06
(E) 3.03

35. The graph of $|y - 1| = |x + 1|$ forms an X. The two branches of the X intersect at a point whose coordinates are

(A) (1,1)
(B) (–1,1)
(C) (1,–1)
(D) (–1,–1)
(E) (0,0)

36. For what value of x between 0° and 360° does cos 2x = 2 cos x?

(A) 68.5° or 291.5°
(B) only 68.5°
(C) 103.9° or 256.1°
(D) 90° or 270°
(E) 111.5° or 248.5°

37. For what value(s) of x will the graph of the function

$f(x) = \sin \sqrt{B - x^2}$ have a maximum?

(A) $\dfrac{\pi}{2}$

(B) $\sqrt{B - \dfrac{\pi}{2}}$

(C) $\sqrt{B - \left(\dfrac{\pi}{2}\right)^2}$

(D) $\pm\sqrt{B - \dfrac{\pi}{2}}$

(E) $\pm\sqrt{B - \left(\dfrac{\pi}{2}\right)^2}$

38. For each positive integer n, let S_n = the sum of all positive integers less than or equal to n. Then S_{51} equals

(A) 50
(B) 51
(C) 1250
(D) 1275
(E) 1326

39. If the graphs of $3x^2 + 4y^2 - 6x + 8y - 5 = 0$ and $(x - 2)^2 = 4(y + 2)$ are drawn on the same coordinate system, at how many points do they intersect?

(A) 0
(B) 1
(C) 2
(D) 3
(E) 4

40. If $\log_x 2 = \log_3 x$ is satisfied by two values of x, what is their sum?

(A) 0
(B) 1.73
(C) 2.35
(D) 2.81
(E) 3.14

41. Which of the following lines are asymptotes for

the graph of $y = \dfrac{3x^2 - 13x - 10}{x^2 - 4x - 5}$

I. $x = -1$

II. $x = 5$

III. $y = 3$

(A) I only
(B) II only
(C) I and II
(D) I and III
(E) I, II, and III

42. If $\dfrac{3\sin 2\theta}{1 - \cos 2\theta} = \dfrac{1}{2}$ and $0° \le \theta \le 180°$, then $\theta =$

(A) 0°
(B) 0° or 180°
(C) 80.5°
(D) 0° or 80.5°
(E) 99.5°

43. If $f(x,y) = 2x^2 - y^2$ and $g(x) = 2^x$, which one of the following is equal to 2^{2x}?

(A) $f(x, g(x))$
(B) $f(g(x), x)$
(C) $f(g(x), g(x))$
(D) $f(g(x), 0)$
(E) $g(f(x,x))$

44. Two positive numbers, a and b, are in the sequence 4, a, b, 12. The first three numbers form a geometric sequence, and the last three numbers form an arithmetic sequence. The difference $b - a$ equals

(A) 1
(B) $1\dfrac{1}{2}$
(C) 2
(D) $2\dfrac{1}{2}$
(E) 3

45. A sector of a circle has an arc length of 2.4 feet and an area of 14.3 square feet. How many degrees are in the central angle?

(A) 63.4°
(B) 20.2°
(C) 14.3°
(D) 12.9°
(E) 11.5°

USE THIS SPACE FOR SCRATCH WORK

$\frac{\theta}{360}\pi r^2 = 14.3$

$\frac{\theta}{360}\pi\left(\frac{432^2}{\theta^2\pi^2}\right) = 14.3$

$\frac{432^2}{360\cdot\theta\cdot\pi} = 14.3$

$\frac{\theta}{360}\cdot 2\pi r = 2.4$

$r = \frac{864}{\theta 2\pi}$

$r = \frac{432}{\theta\pi}$

46. The *y*-coordinate of one focus of the ellipse $36x^2 + 25y^2 + 144x - 50y - 731 = 0$ is

(A) –2
(B) 1
(C) 3.32
(D) 4.32
(E) 7.81

$36x^2 + 144x \underline{\quad} + 25y^2 - 50y \underline{\quad} = 731$

$36(x^2 + 4x \underline{+4}) + 25(y^2 - 2y \underline{+1}) = 731 + 144 + 25$

$36(x+2)^2 + 25(y-1)^2 = 900$

$\frac{(x+2)^2}{25} + \frac{(y-1)^2}{36} = 1$

$36 = 25 + c^2$

$11 = c^2$

$c = \sqrt{11}$

47. In the figure above, *ABCD* is a square. *M* is the point one-third of the way from *B* to *C*. *N* is the point one-half of the way from *D* to *C*. Then θ =

(A) 50.8°
(B) 45.0°
(C) 36.9°
(D) 36.1°
(E) 30.0°

48. If *f* is a linear function such that *f*(7) = 5, *f*(12) = –6, and *f*(*x*) = 23.7, what is the value of *x*?

(A) –3.2
(B) –1.5
(C) 1
(D) 2.4
(E) 3.1

7, 5

12, –6

$m = \frac{5+6}{7-12} = \frac{11}{-5}$

$y - 5 = -\frac{11}{5}(x-7)$

$23.7 - 5 = -\frac{11}{5}(x-7)$

49. Under which of the following conditions is $\dfrac{x(x-y)}{y}$ negative?

    (A) $x < y < 0$
    (B) $y < x < 0$
    (C) $0 < y < x$
    (D) $x < 0 < y$
    (E) All of the above.

50. The binary operation $*$ is defined over the set of real numbers to be $a * b = \begin{cases} a\sin\dfrac{b}{a} & \text{if } a > b \\ b\cos\dfrac{a}{b} & \text{if } a < b \end{cases}$.

    What is the value of $2 * (5 * 3)$?

    (A) 1.84
    (B) 2.14
    (C) 2.79
    (D) 3.65
    (E) 4.01

USE THIS SPACE FOR SCRATCH WORK

STOP

If there is still time remaining, you may review your answers.

Answer Key
MODEL TEST 4

1. **D**	14. **B**	27. **C**	40. **D**
2. **A**	15. **D**	28. **D**	41. **D**
3. **C**	16. **D**	29. **D**	42. **C**
4. **B**	17. **B**	30. **E**	43. **C**
5. **A**	18. **E**	31. **A**	44. **E**
6. **A**	19. **C**	32. **E**	45. **E**
7. **A**	20. **C**	33. **A**	46. **D**
8. **B**	21. **D**	34. **E**	47. **B**
9. **A**	22. **E**	35. **B**	48. ~~D~~ B
10. **B**	23. **D**	36. **E**	49. **A**
11. **D**	24. **C**	37. **E**	50. **B**
12. **B**	25. **D**	38. **E**	
13. **B**	26. **B**	39. **C**	

ANSWERS EXPLAINED

The following explanations are keyed to the review portions of this book. The number in brackets after each explanation indicates the appropriate section in the Review of Major Topics (Part 2). If a problem can be solved using algebraic techniques alone, [algebra] appears after the explanation, and no reference is given for that problem in the Self-Evaluation Chart at the end of the test.

In these solutions the following notation is used:

i: calculator unnecessary
a: calculator helpful or necessary
g: graphing calculator helpful or necessary

1. i **(D)** Since inverse functions are symmetric about the line $y = x$, if point (a,b) lies on f, point (b,a) must lie on f^{-1}. [1.1]

2. a **(A)** Average $= \dfrac{70+80+85+80+x}{5} = 80$. Therefore, $x = 85$. [4.1]

3. g **(C)** Plot the graph $y = (x^2 + 3x + 2)/(x + 2)\tan(\pi x)$ in a $[-2.5,2.5]$ by $[-5,5]$ window and use TRACE to approximate the location of asymptotes. The only answer choice that could be an asymptote occurs at $x = \dfrac{1}{2}$.

   An alternative solution is to factor the numerator of $f(x)$ and observe that the factor $x + 2$ divides out: $f(x) = \dfrac{(x+2)(x+1)}{x+2} \tan \pi x = (x+1)\tan \pi x$. The only asymptotes occur because of $\tan \pi x$, when πx is a multiple of $\dfrac{\pi}{2}$. Setting $\pi x = \dfrac{\pi}{2}$ yields $x = \dfrac{1}{2}$. [1.2, 1.3]

4. i **(B)** The bases are the same, so the log of a product equals the sum of the logs. [1.4]

5. g **(A)** Plot the graph of $y = 3x^3 + 4x^2 - 4x$ in the standard window and use CALC/zero to find the three zeros of this function. Sum these three values to get the correct answer choice.

   An alternative solution is first to observe that x factors out, so that $x = 0$ is one zero. The other factor is a quadratic, so the sum of its zeros is $-\dfrac{b}{a} = -\dfrac{4}{3}$. [1.2]

6. i **(A)** $f(a) = a - \dfrac{1}{a}$ and $f\left(\dfrac{1}{a}\right) = \dfrac{1}{a} - a$. Therefore, $f(a) + f\left(\dfrac{1}{a}\right) = 0$. [1.1]

7. a **(A)** Enter the function $f(x)$ into Y_1 and return to the Home Screen. Enter $Y_1(2) - Y_1\left(\sqrt{2}\right)$ to get the correct answer choice.

   An alternative solution is to use a scientific calculator to evaluate the function at $x = 2$ and at $x = \sqrt{2}$, and then subtract. [1.1]

8. i **(B)** The $f(2 - x)$ just shifts and reflects the graph horizontally; it does not have any vertical effect on the graph. Therefore, regardless of what is substituted for x, $f(x) \geq 0$. [2.1]

9. i (A) Only 3 of the numbers can be used in the thousands place, 3 are left for the hundreds place, 2 for the tens place, and only one for the units place. $3 \cdot 3 \cdot 2 \cdot 1 = 18$. [3.1]

10. i (B) Substituting 1 for x gives $1 + a - 4 = 0$, and so $a = 3$. [1.2]

11. a (D) The first 5 flips have no effect on the next 5 flips, so the problem becomes "What is the probability of getting exactly 3 heads in 5 flips of a coin?" $\binom{5}{3} = 10$ outcomes contain 3 heads out of a total of $2^5 = 32$ possible outcomes. $P(3\text{H}) = \dfrac{5}{16} \approx 0.3125$. [4.2]

12. i (B) $i^{4n} = 1$; $i^{4n+1} = i$; $i^{4n+2} = -1$; $i^{4n+3} = -i$; $i^{4n+4} = (i^{4n})(i^4) = (1)(1)$. [3.2]

13. i (B) Since $\sqrt{x} = x^{1/2}$, $\log_r \sqrt{3} = \log_r 3^{1/2} = \dfrac{1}{2} \log_r 3 = 3.55$. [1.4]

14. i (B) The 23° is superfluous because $g(x) = f(\sin x) + f(\cos x) = 4\sin^2 x + 4\cos^2 x = 4(\sin^2 x + \cos^2 x) = 4(1) = 4$. [1.1, 1.3]

15. a (D) This is a tricky problem. If you just plot the graph of $y = \left(x - \sqrt{2}\right)\left(x^2 - \sqrt{3}x + \pi\right)$, you will see only one real zero, at approximately 1.414 ($\approx \sqrt{2}$). This is because the zeros of the quadratic factor are imaginary. Since, however, they are imaginary conjugates, their sum is real—namely twice the real part. Therefore, graphing the function, using CALC/zero to find the zeros, and summing them will give you the wrong answer.

To get the correct answer, you must use the fact that the sum of the zeros of the quadratic factor is $-\dfrac{b}{a} = \sqrt{3} \approx 1.732$. Since the zero of the linear factor is $\sqrt{2} \approx 1.414$, the sum of the zeros is about $1.732 + 1.414 \approx 3.15$. [1.2]

16. i (D) Graph I consists of the lines $x = 0$ and $y = 0$, which are the coordinate axes and are therefore perpendicular. Graph II consists of $y = |x|$ and $y = -|x|$, which are at ±45° to the coordinate axes and are therefore perpendicular. Graph III consists of the hyperbolas $xy = 1$ and $xy = -1$. Therefore, the correct answer choice is D.

There are two reasons why a graphing calculator solution is not recommended here. One is that equations, not functions, are given, and solving these equations so that they can be graphed involves two branches each. The other reason is that even with graphs, you would have to make judgments about perpendicularity. At a minimum, this would require you to graph the equations in a square window. [1.6]

17. i (B) Points 6 in. from m form a cylinder, with m as axis, which is tangent to plane X. Points 1 in. from X are two planes parallel to X, one above and one below X. The cylinder intersects only one of the planes in two lines parallel to m. [2.2]

18. i (E) Since the y-coordinate of the point V could be at any height, the slope of VO could be any value. [2.1]

19. a (C) Volume of cylinder $= \pi r^2 h = 36\pi =$ volume of cube $= s^3$. Therefore, $s = \sqrt[3]{36\pi} \approx 4.84$. [2.2]

20. i (C) Substitute $2t - 2$ for x. [1.6]

21. a (D) $y_1 = 2^{3/2} + 2a + \sqrt{2}b + c$ and $y_2 = -(2)^{3/2} + 2a - 2\sqrt{2}b + c$. So, $y_1 - y_2 =$

$(2^{3/2} + 2^{3/2}) + 2\sqrt{2}b = 3$. Therefore, $5.65685 + 2.828b \approx 3$ and $b \approx \dfrac{3 - 5.65685}{2.8284} \approx$ -0.939. [1.2]

22. a (E) The probability that the first car selected is compact is $\dfrac{20}{27}$. There are 26 cars left, of which 19 are compact. The probability that the second car is also compact is $\left(\dfrac{20}{27}\right) \cdot \left(\dfrac{19}{26}\right) \approx 0.541$. [4.2]

23. i (D) Here, a could be either positive or negative. However, b must be negative. [algebra]

24. g (C) Plot the graph of $y = \text{abs}(x - \text{int}(x) - 1/2)$ in an $x\varepsilon[-5,5]$ and $y\varepsilon[-2,2]$ window and observe that the maximum value is $\dfrac{1}{2}$.

An alternative solution is to sketch a portion of the graph by hand and observe the maximum value. [1.6]

25. g (D) Plot the graph $y = (x^3 - 8)(x^2 - 4)$ in the standard window. Using CALC/value, observe that y is not defined when $x = 2$. Therefore, enter 1.999 for x and observe that y is approximately equal to 3.

An alternative solution is to factor the numerator and denominator, divide out $x - 2$, and substitute the limiting value 2 into the resulting expression:

$$\lim_{x \to 2} \frac{x^3 - 8}{x^2 - 4} = \lim_{x \to 2} \frac{(x - 2)(x^2 + 2x + 2)}{(x - 2)(x + 2)}$$

$$= \lim_{x \to 2} \frac{(x^2 + 2x + 4)}{x + 2} = \frac{4 + 4 + 4}{2 + 2} = 3. \quad [1.5]$$

26. a (B) A sketch will help you see that the height of the cone is $5 + 3 = 8$.

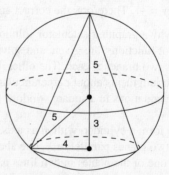

The volume of the cone is $V_c = \dfrac{1}{3}\pi r^2 h = \dfrac{1}{3}\pi(16)(8)$, and the volume of the sphere is

$V_s = \dfrac{4}{3}\pi r^3 = \dfrac{4}{3}\pi(125)$. The desired ratio is $V_c : V_s = 0.256 : 1$. [2.2]

27. g (C) Enter 1 into your calculator. Then enter $\sqrt[3]{2\text{Ans}}$ three times, to accomplish three iterations that result in x_3 to get the correct answer choice E.

An alternative solution is to use the formula to evaluate x_1, x_2, and x_3, in turn. [3.4]

28. g (D) With your calculator in radian mode, plot the graph of $y = \text{abs}\left(\sqrt{2}\left(1/\sin\left(x + \pi/5\right)\right)\right)$ in an $x\varepsilon[-1,1]$ and $y\varepsilon[0,5]$ window. Use VALUE/X = 0 to determine that the y-intercept is approximately 1.49. [1.3]

29. i (D) The equation of the circle is $(x - 4)^2 + (y + 8)^2 = r^2$. Multiplying out indicates that $a = -8$ and $b = 16$, and so $a + b = 8$. [2.1]

30. g (E) Since $p(a) = 2$, $3a^2 + 9a + 5 = 0$. Solve this quadratic by using your program for the Quadratic Formula to get the correct answer choice.

An alternative solution is to substitute 3 for a, 9 for b, and 5 for c in the Quadratic Formula and evaluate. [1.2]

31. g (A) Since i is a root of the equation, so is $-i$, so there are 2 real roots. Plot the graph of $y = x^4 + 2x^3 - 3x^2 + 2x - 4$ in the standard window. Use CALC/zero to find one of the roots, and store the answer (X) in A. Then use CALC/zero again to find the other root, and multiply A and X to get the correct answer choice. [3.2]

32. g (E) First find $\sin^{-1}(3/5) \approx 36.87°$, the reference angle for angle A. Since A is in the second quadrant, $A \approx 180° - 36.87° = 143.13°$. Store 180-Ans in A. Similarly, find $\cos^{-1}(1/3) \approx 70.53°$, the reference angle for angle B. Since B is in the fourth quadrant, store 360-Ans in B. Then evaluate $\sin(A + B)$ to get the correct answer choice. [1.3]

33. a (A) The probability that at least one is a boy is (1 – the probability that all 3 are girls) or $1 - (0.5)^3 = 0.875$. [4.2]

34. a (E) Since $\sec 1.4 = x$, $\cos 1.4 = \dfrac{1}{x}$ or $x = \dfrac{1}{\cos 1.4}$. Therefore, $\csc(2\,\text{Arctan}\,x) =$ $1/\sin(2\tan^{-1}(1/\cos 1.4)) \approx 3.03$. [1.3]

35. g (B) The equation $|y - 1| = |x + 1|$ defines two functions: $y = \pm|x + 1| + 1$. Plot these graphs in the standard window and observe that they intersect at $(-1,1)$.

An alternative solution is to recall that the important point of an absolute value occurs when the expression within the absolute value sign equals zero. The important point of this absolute value problem occurs when $y - 1 = 0$ and $x + 1 = 0$, i.e., at $(-1,1)$. [1.6]

36. g (E) With your calculator in degree mode, plot the graphs of $y = \cos 2x$ and $y = 2\cos x$ in an $x\varepsilon[0,360]$ and $y\varepsilon[-2,2]$ window. Use CALC/intersect to find the correct answer choice E. [1.3]

37. i (E) Since $\sin\theta$ has a maximum at $\theta = \dfrac{\pi}{2}$, $\sqrt{B - x^2} = \dfrac{\pi}{2}$. Thus, $B - x^2 = \left(\dfrac{\pi}{2}\right)^2$ and

$x^2 = B - \left(\dfrac{\pi}{2}\right)^2$. Therefore, $x = \pm\sqrt{B - \left(\dfrac{\pi}{2}\right)^2}$. [1.3]

38. g (E) Enter LIST/MATH/sum(LIST/OPS/seq(X,X,1,51)) to compute the desired sum.

An alternative solution is to observe that the sequence is arithmetic with $t_1 = 1$ and $d = 1$. Using the formula for the sum of the first n terms of an arithmetic sequence,

$$S_{51} = \frac{51}{2}(2 + 50 \cdot 1) = 51 \cdot 26 = 1326. \ [3.4]$$

39. g (C) Complete the square in the first equation to get $3(x-1)^2 + 4(y+1)^2 = 12$. Solving

this equation for y yields $y = \pm\sqrt{\dfrac{12 - 3(x-1)^2}{4}} - 1$. Solving for y in the second

equation, $y = \dfrac{(x-2)^2}{4} - 2$. Plot the graphs of these three equations in the standard window to see that the graphs intersect in two places.

An alternative solution can be found by completing the square in the first equation and

dividing by 12 to get the standard form equation of an ellipse, $\dfrac{(x-1)^2}{4} + \dfrac{(y+1)^2}{3} = 1$.

The second equation is the standard form of a parabola. Sketch these two equations and observe the number of points of intersection. [2.1]

40. a (D) Let $y = \log_x 2 = \log_3 x$. Converting to exponential form gives $x^y = 2$ and $3^y = x$.

Substitute to get $3^{y^2} = 2$, which can be converted into $y^2 = \dfrac{\log 2}{\log 3} \approx 0.6309$. Thus,

$y \approx \pm 0.7943$. Therefore, $3^{0.7943} = x \approx 2.393$ or $3^{-0.7943} = x \approx 0.4178$. Therefore, the sum of two x's is 2.81. [1.4]

41. i (D) Factor the numerator and denominator: $\dfrac{(3x+2)(x-5)}{(x+1)(x-5)}$. Since the $x - 5$ divide

out, the only vertical asymptote is at $x = -1$. Since the degree of the numerator and denominator are equal, y approaches 3 as x approaches $\pm\infty$, so $y = 3$ is a horizontal asymptote. [1.5]

42. g (C) With your calculator in degree mode, plot the graphs of $y = 1/2$ and $y = (3\sin(2x))/(1 - \cos(2x))$ in the window [0,180] by [-2,2] Use CALC/intersect to find the one point of intersection in the specified interval, at 80.5°.

An alternative solution is to cross-multiply the original equation and use the double angle formulas for sine and cosine, to get

$6\sin 2\theta = 1 - \cos 2\theta$

$12\sin\theta\cos\theta = 1 - (1 - 2\sin^2\theta) = 2\sin^2\theta$

$6\sin\theta\cos\theta - \sin^2\theta = 0$

$\sin\theta(6\cos\theta - \sin\theta) = 0$

Therefore, $\sin\theta = 0$ or $\tan\theta = 6$. It follows that $\theta = 0°, 180°$, or 80.5°. The first two solutions make the denominator of the original equation equal to zero, so 80.5° is the only solution. [1.3]

43. i (C) Backsolve until you get $f(g(x), g(x)) = 2(2^x)^2 - (2^x)^2 = (2^x)^2 = 2^{2x}$. [1.1]

44. i (E) From the geometric sequence, $b = a\left(\dfrac{a}{4}\right)$. From the arithmetic sequence,

$b - a = 12 - b$, or $2b - a = 12$. Substituting gives $2\left(\dfrac{a^2}{4}\right) - a = 12$.

Solving gives $a = 6$ or -4. Eliminate -4 since a is given to be positive. Substituting the 6 gives $2b - 6 = 12$, giving $b = 9$. Therefore, $b - a = 3$. [3.4]

45. a (E) Since $s = r\theta^R$, then $2.4 = r\theta$, which implies that $r = \dfrac{2.4}{\theta}$. $A = \dfrac{1}{2}r^2\theta$, and so

$14.3 = \dfrac{1}{2}r^2\theta$, which implies that $r^2 = \dfrac{28.6}{\theta}$. Therefore, $\left(\dfrac{2.4}{\theta}\right)^2 = \dfrac{28.6}{\theta}$, which

implies that $2.4^2 = 28.6\theta$. Therefore, $\theta = \dfrac{5.76}{28.6} \approx 0.2014^R \approx 11.5°$. [1.3]

46. a (D) Complete the square and put the equation of the ellipse in standard form:

$36x^2 + 25y^2 + 144x - 50y - 731$

$\quad = 36(x+2)^2 + 25(y-1)^2 - 900$

$\dfrac{(x+2)^2}{25} + \dfrac{(y-1)^2}{36} = 1$

The center of the ellipse is at $(-2,1)$, with $a^2 = 36$ and $b^2 = 25$, and the major axis is parallel to the y axis. Each focus is $c = \sqrt{a^2 - b^2}$ units above and below the center.

Therefore, the y coordinates of the foci are $1 \pm \sqrt{11} \approx 4.32$ and -2.32. [2.1]

47. a (B) Because you are bisecting one side and trisecting another side, it is convenient to let the length of the sides be a number divisible by both 2 and 3. Let $AB = AD = 6$. Thus $BM = 2$, $MC = 4$, and $CN = ND = 3$. Let $\angle NAD = x$, so that, using right triangle

NAD, $\tan x = \dfrac{3}{6} = 0.5$, which implies that $x = \tan^{-1} 0.5 \approx 26.6°$. Let $\angle MAB = y$, so

that, using right triangle MAB, $\tan y = \dfrac{2}{6}$, which implies that $y = \tan^{-1}\dfrac{1}{3} \approx 18.4°$.

Therefore, $\theta \approx 90° - 26.6° - 18.4° \approx 45°$. [1.3]

48. a (B) The slope of the line is $\dfrac{-6 - 5}{12 - 7} = -\dfrac{11}{5}$. An equation of the line is therefore

$y - 5 = -\dfrac{11}{5}(x - 7)$. Substitute 23.7 for y and solve for x to get $x = -1.5$. [1.2]

49. i (A) You must check each answer choice, one at a time. In A, $x < 0$, $y < 0$, $x - y < 0$, so the expression is negative. In B, $x < 0$, $y < 0$, $x - y > 0$, so the expression is positive. At this point you know that the correct answer choice must be A. [algebra]

50. a (B) Put your calculator in radian mode: $5 * 3 = 5\sin\dfrac{3}{5} \approx 2.823$.

$2 * (5 * 3) = 2 * 2.823 \approx 2.823 \cos\dfrac{2}{2.823} \approx 2.14$. [1.1]

Test 4–Self-Evaluation Chart

Self-Evaluation Chart for Model Test 4

Subject Area	Questions and Review Section							Right	Number Wrong	Omitted
Algebra and Functions (24 questions)	1 1.1	3 1.2	4 1.4	5 1.2	6 1.1	7 1.1		6	0	0
	10 1.2	13 1.4	14 1.1	15ʷ 1.2	16ʷ 1.6	20 1.6	21 1.2	5	2	0
	23	24 1.6	25ʷ 1.5	30 1.2	35 1.6	40 1.4		5	1	0
	41 1.5	43 1.1	48° 1.2	49°	50° 1.1			2	0	3
Trigonometry (11 questions)	3 1.3	14 1.3	28 1.3	32 1.3	34 1.3	36 1.3		6	0	0
	37 1.3	40 1.4	42ʷ 1.3	45 1.3	47° 1.3			3	1	1
Coordinate and Three-Dimensional Geometry (8 questions)	8 2.1	17ʷ 2.2	18 2.1	19 2.2	26ʷ 2.2			3	2	0
	29 2.1	39° 2.1	46° 2.1					1	0	2
Numbers and Operations (6 questions)	9ʷ 3.1	12 3.2	27ʷ 3.4	31° 3.2	38 3.4	44 3.4		3	2	1
Data Analysis, Statistics, and Probability (4 questions)	2 4.2	11° 4.2	22 4.2	33 4.2				3	0	1
TOTALS								37	8	8

Evaluate Your Performance Model Test 4	
Rating	**Number Right**
Excellent	41–50
Very good	33–40
Above average	25–32
Average	15–24
Below average	Below 15

Calculating Your Score

Raw score R = number right $- \frac{1}{4}$ (number wrong), rounded = _____ 35

Approximate scaled score $S = 800 - 10(44 - R) =$ _____ 710

If $R \geq 44$, $S = 800$.

Answer Sheet

MODEL TEST 5

1 Ⓐ Ⓑ Ⓒ Ⓓ	14 Ⓐ Ⓑ Ⓒ Ⓓ	27 Ⓐ Ⓑ Ⓒ Ⓓ	40 Ⓐ Ⓑ Ⓒ Ⓓ
2 Ⓐ Ⓑ Ⓒ Ⓓ	15 Ⓐ Ⓑ Ⓒ Ⓓ	28 Ⓐ Ⓑ Ⓒ Ⓓ	41 Ⓐ Ⓑ Ⓒ Ⓓ
3 Ⓐ Ⓑ Ⓒ Ⓓ	16 Ⓐ Ⓑ Ⓒ Ⓓ	29 Ⓐ Ⓑ Ⓒ Ⓓ	42 Ⓐ Ⓑ Ⓒ Ⓓ
4 Ⓐ Ⓑ Ⓒ Ⓓ	17 Ⓐ Ⓑ Ⓒ Ⓓ	30 Ⓐ Ⓑ Ⓒ Ⓓ	43 Ⓐ Ⓑ Ⓒ Ⓓ
5 Ⓐ Ⓑ Ⓒ Ⓓ	18 Ⓐ Ⓑ Ⓒ Ⓓ	31 Ⓐ Ⓑ Ⓒ Ⓓ	44 Ⓐ Ⓑ Ⓒ Ⓓ
6 Ⓐ Ⓑ Ⓒ Ⓓ	19 Ⓐ Ⓑ Ⓒ Ⓓ	32 Ⓐ Ⓑ Ⓒ Ⓓ	45 Ⓐ Ⓑ Ⓒ Ⓓ
7 Ⓐ Ⓑ Ⓒ Ⓓ	20 Ⓐ Ⓑ Ⓒ Ⓓ	33 Ⓐ Ⓑ Ⓒ Ⓓ	46 Ⓐ Ⓑ Ⓒ Ⓓ
8 Ⓐ Ⓑ Ⓒ Ⓓ	21 Ⓐ Ⓑ Ⓒ Ⓓ	34 Ⓐ Ⓑ Ⓒ Ⓓ	47 Ⓐ Ⓑ Ⓒ Ⓓ
9 Ⓐ Ⓑ Ⓒ Ⓓ	22 Ⓐ Ⓑ Ⓒ Ⓓ	35 Ⓐ Ⓑ Ⓒ Ⓓ	48 Ⓐ Ⓑ Ⓒ Ⓓ
10 Ⓐ Ⓑ Ⓒ Ⓓ	23 Ⓐ Ⓑ Ⓒ Ⓓ	36 Ⓐ Ⓑ Ⓒ Ⓓ	49 Ⓐ Ⓑ Ⓒ Ⓓ
11 Ⓐ Ⓑ Ⓒ Ⓓ	24 Ⓐ Ⓑ Ⓒ Ⓓ	37 Ⓐ Ⓑ Ⓒ Ⓓ	50 Ⓐ Ⓑ Ⓒ Ⓓ
12 Ⓐ Ⓑ Ⓒ Ⓓ	25 Ⓐ Ⓑ Ⓒ Ⓓ	38 Ⓐ Ⓑ Ⓒ Ⓓ	
13 Ⓐ Ⓑ Ⓒ Ⓓ	26 Ⓐ Ⓑ Ⓒ Ⓓ	39 Ⓐ Ⓑ Ⓒ Ⓓ	

Model Test 5

T ear out the preceding answer sheet. Decide which is the best choice by rounding your answer when appropriate. Blacken the corresponding space on the answer sheet. When finished, check your answers with those at the end of the test. For questions that you got wrong, note the sections containing the material that you must review. Also if you do not fully understand how you arrived at some of the correct answers, you should review the appropriate sections. Finally, fill out the self-evaluation chart on page 325 in order to pinpoint the topics that give you the most difficulty.

50 questions: 1 hour

Directions: Decide which answer choice is best. If the exact numerical value is not one of the answer choices, select the closest approximation. Fill in the oval on the answer sheet that corresponds to your choice.

Notes:
(1) You will need to use a scientific or graphing calculator to answer some of the questions.
(2) You will have to decide whether to put your calculator in degree or radian mode for some problems.
(3) All figures that accompany problems are plane figures unless otherwise stated. Figures are drawn as accurately as possible to provide useful information for solving the problem, except when it is stated in a particular problem that the figure is not drawn to scale.
(4) Unless otherwise indicated, the domain of a function is the set of all real numbers for which the functional value is also a real number.

Reference Information. The following formulas are provided for your information.

Volume of a right circular cone with radius r and height h: $V = \frac{1}{3}\pi r^2 h$

Lateral area of a right circular cone if the base has circumference c and slant height is l:

$S = \frac{1}{2}cl$

Volume of a sphere of radius r: $V = \frac{4}{3}\pi r^3$

Surface area of a sphere of radius r: $S = 4\pi r^2$

Volume of a pyramid of base area B and height h: $V = \frac{1}{3}Bh$

1. $x^{2/3} + x^{4/3} =$

   (A) $x^{2/3}$
   (B) $x^{8/9}$
   (C) x
   (D) x^2
   (E) $x^{2/3}(x^{2/3} + 1)$

2. In three dimensions, what is the set of all points for which $x = 0$?

   (A) the origin
   (B) a line parallel to the x-axis
   (C) the yz-plane
   (D) a plane containing the x-axis
   (E) the x-axis

3. Expressed with positive exponents only, $\dfrac{ab^{-1}}{a^{-1} - b^{-1}}$

   is equivalent to

   (A) $\dfrac{a^2}{a - b}$

   (B) $\dfrac{a^2}{a - 1}$

   (C) $\dfrac{b - a}{ab}$

   (D) $\dfrac{a^2}{b - a}$

   (E) $\dfrac{1}{a - b}$

4. If $f(x) = \sqrt[3]{x}$ and $g(x) = x^3 + 8$, find $(f \circ g)(3)$.

   (A) 3.3
   (B) 5
   (C) 11
   (D) 35
   (E) 50.5

5. $x > \sin x$ for

   (A) all $x > 0$
   (B) all $x < 0$
   (C) all x for which $x \neq 0$
   (D) all x
   (E) all x for which $-\dfrac{\pi}{2} < x < 0$

6. The sum of the zeros of $f(x) = 3x^2 - 5$ is

   (A) 3.3
   (B) 1.8
   (C) 1.7
   (D) 1.3
   (E) 0

7. The intersection of a plane with a right circular cylinder could be which of the following?

   I. A circle ✓
  II. Parallel lines
 III. Intersecting lines

(A) I only
(B) II only
(C) III only
(D) I and II only
(E) I, II, and III

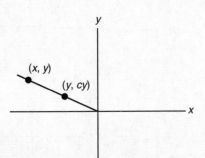

8. In the figure, c equals

(A) 1
(B) xy
(C) $\dfrac{x}{y}$
(D) $\dfrac{y}{x}$
(E) -1

$\dfrac{y}{x} = \dfrac{cy}{y}$

9. The graph of $f(x) = \dfrac{10}{x^2 - 10x + 25}$ has a vertical asymptote at $x =$

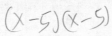
$(x-5)(x-5)$

(A) 0 only
(B) 5 only
(C) 10 only
(D) 0 and 5 only
(E) 0, 5, and 10

10. $P(x) = x^5 + x^4 - 2x^3 - x - 1$ has at most n positive zeros. Then $n =$

(A) 0
(B) 1
(C) 2
(D) 3
(E) 5

11. If x and y are real numbers and $i = \sqrt{-1}$, find the ordered pair solution for the equation $3yi + 2i^3 = 7i + 6i^6 + 3x$.

(A) $\left(0, \dfrac{7}{3}\right)$

(B) $(-3,3)$

(C) $(2,3)$

(D) $\left(\dfrac{3}{2}, \dfrac{3}{7}\right)$

(E) $(3,2)$

12. If $f(x)$ is a linear function and $f(2) = 1$ and $f(4) = -2$, then $f(x) =$

(A) $-\dfrac{3}{2}x + 4$

(B) $\dfrac{3}{2}x - 2$

(C) $-\dfrac{3}{2}x + 2$

(D) $\dfrac{3}{2}x - 4$

(E) $-\dfrac{2}{3}x + \dfrac{7}{3}$

13. The length of the radius of a circle is one-half the length of an arc of the circle. How large is the central angle that intercepts that arc?

(A) $60°$

(B) $120°$

(C) 1^R

(D) 2^R

(E) π^R

14. If $f(x) = 2^x + 1$, then $f^{-1}(7) =$

(A) 2.4

(B) 2.6

(C) 2.8

(D) 3

(E) 3.6

15. Find all values of x that satisfy the determinant equation $\begin{vmatrix} 2x & 1 \\ x & x \end{vmatrix} = 3$.

(A) -1

(B) -1 or 1.5

(C) 1.5

(D) -1.5

(E) -1.5 or 1

16. The 71st term of 30, 27, 24, 21, . . . , is

    (A) 5325
    (B) 240
    (C) 180
    (D) −180
    (E) −183

Handwritten: $d = -3$

Handwritten: $a_n = 30 + (n-1)(-3)$

17. If $0 < x < \dfrac{\pi}{2}$ and $\tan 5x = 3$, to the nearest tenth,

    what is the value of $\tan x$?

    (A) 0.5
    (B) 0.4
    (C) 0.3
    (D) 0.2
    (E) 0.1

18. If $4.05^p = 5.25^q$, what is the value of $\dfrac{p}{q}$?

    (A) −0.11
    (B) 0.11
    (C) 1.19
    (D) 1.30
    (E) 1.67

Handwritten: $p \log 4.05 = q \log 5.25$

Handwritten: $\dfrac{p}{q} = \dfrac{\log 5.25}{\log 4.05}$

19. A cylinder has a base radius of 2 and a height of 9.
    To the nearest whole number, by how much does the
    lateral area exceed the sum of the areas of the two
    bases?

    (A) 101
    (B) 96
    (C) 88
    (D) 81
    (E) 75

Handwritten: $C = 2\pi 2 = 4\pi$

Handwritten: $A_{base} = 2\pi(2)^2$

20. If $\cos 67° = \tan x°$, then $x =$

    (A) 0.4
    (B) 6.8
    (C) 7.8
    (D) 21
    (E) 29.3

21. $P(x) = x^3 + 18x - 30$ has a zero in the interval

    (A) (0, 0.5)
    (B) (0.5, 1)
    (C) (1, 1.5)
    (D) (1.5, 2)
    (E) (2, 2.5)

22. The lengths of the sides of a triangle are 23, 32, and 37. To the nearest degree, what is the value of the largest angle?

    (A) 71°
    (B) 83°
    (C) 122°
    (D) 128°
    (E) 142°

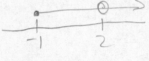

$37^2 = 32^2 + 23^2 - 2 \cdot 32 \cdot 23 \cos\theta$

23. If $f(x) = \dfrac{3}{x-2}$ and $g(x) = \sqrt{x+1}$, find the domain of $f \circ g$.

    $f : X \neq 2$ $\quad g : x \geq -1$

    (A) $x \geq -1$
    (B) $x \neq 2$
    (C) $x \geq -1, x \neq 2$
    (D) $x \geq -1, x \neq 3$
    (E) $x \leq -1$

24. Two cards are drawn from a regular deck of 52 cards. What is the probability that both will be 7s?

    (A) 0.149
    (B) 0.04
    (C) 0.012
    (D) 0.009
    (E) 0.005

    $\dfrac{4}{52} \cdot \dfrac{3}{51}$

25. If $\sqrt{y} = 3.216$, then $\sqrt{10y} =$

    (A) 321.6
    (B) 32.16
    (C) 10.17
    (D) 5.67
    (E) 4.23

26. What is the domain of the function

    $f(x) = \log \sqrt{2x^2 - 15}$?

    (A) $-7.5 < x < 7.5$
    (B) $x < -7.5$ or $x > 7.5$
    (C) $x < -2.7$ or $x > 2.7$
    (D) $x < -3.2$ or $x > 3.2$
    (E) $x < 1.9$ or $x > 1.9$

    $2x^2 - 15 > 0$
    $2x^2 > 15$
    $x^2 > \dfrac{15}{2}$

    $x > \pm 7.5$

27. A magazine has 1,200,000 subscribers, of whom 400,000 are women and 800,000 are men. Twenty percent of the women and 60 percent of the men read the advertisements in the magazine. What is the probability that a randomly selected subscriber reads the advertisements?

    (A) 0.30
    (B) 0.36
    (C) 0.40
    (D) 0.47
    (E) 0.52

28. Let S be the sum of the first n terms of the arithmetic sequence 3, 7, 11, . . . , and let T be the sum of the first n terms of the arithmetic sequence 8, 10, 12, ... For $n > 1$, $S = T$ for

    (A) no value of n
    (B) one value of n
    (C) two values of n
    (D) three values of n
    (E) four values of n

29. On the interval $\left[-\dfrac{\pi}{4}, \dfrac{\pi}{4}\right]$, the function

    $f(x) = \sqrt{1 + \sin^2 x}$ has a maximum value of

    (A) 0.78
    (B) 1
    (C) 1.1
    (D) 1.2
    (E) 1.4

30. A point has rectangular coordinates (3,4). The polar coordinates are (5,θ). What is the value of θ?

    (A) 30°
    (B) 37°
    (C) 51°
    (D) 53°
    (E) 60°

31. If $f(x) = x^2 - 4$, for what real number values of x will $f(f(x)) = 0$?

    (A) 2.4
    (B) ±2.4
    (C) 2 or 6
    (D) ±1.4 or ±2.4
    (E) no values

USE THIS SPACE FOR SCRATCH WORK

32. If $f(x) = x \log x$ and $g(x) = 10^x$, then $g(f(2)) =$

    (A) 24
    (B) 17
    (C) 4
    (D) 2
    (E) 0.6

    [handwritten: $f(2) = 2\log 2$]
    [handwritten: $10^{2\log 2} = 10^{\log 4}$]

33. If $f(x) = x^{\sqrt{x}}$, then $f(\sqrt{2}) =$

    (A) 1.4
    (B) 1.5
    (C) 1.6
    (D) 2.0
    (E) 2.7

    [handwritten: $\sqrt{\sqrt{2}}$]
    [handwritten: $\sqrt{2}$]

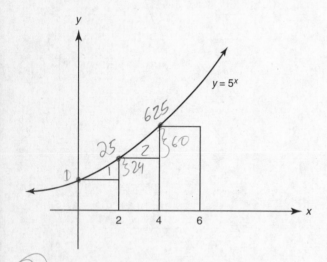

34. The figure above shows the graph of 5^x. What is the sum of the areas of the triangles?

    (A) 32,550
    (B) 16,225
    (C) 2604
    (D) 1302
    (E) 651

    [handwritten: $A_1 = \frac{1}{2} \cdot 2 \cdot 24 = 24$]
    [handwritten: $A_2 = \frac{1}{2} \cdot 2 \cdot 600 = 600$]

35. (p,q) is called a *lattice point* if p and q are both integers. How many lattice points lie in the area between the two curves $x^2 + y^2 = 9$ and $x^2 + y^2 - 6x + 5 = 0$?

    (A) 0
    (B) 1
    (C) 2
    (D) 3
    (E) 4

    [handwritten: $x^2 - 6x + 9 + y^2 + 5 = 0 + 9$]
    [handwritten: $(x-3)^2 + y^2 = 4$]
    [handwritten: $\frac{(x-3)^2}{4} + \frac{y^2}{4} = 1$]
    [handwritten: $y = 1.8$]
    [handwritten: $x = 2\frac{1}{3}$]

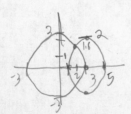

36. If $\sin A = \dfrac{3}{5}$, $90° < A < 180°$, $\cos B = \dfrac{1}{3}$, and

$270° < B < 360°$, the value of $\sin (A + B)$ is
    (A) –0.83
    (B) –0.55

    (C) –0.33
    (D) 0.73
    (E) 0.95

37. For all real numbers x, $f(2x) = x^2 – x + 3$. An expression for $f(x)$ in terms of x is

    (A) $2x^2 – 2x + 3$

    (B) $4x^2 – 2x + 3$

    (C) $\dfrac{x^2}{4} – \dfrac{x}{2} + 3$

    (D) $\dfrac{x^2}{2} – \dfrac{x}{2} + 3$

    (E) $x^2 – x + 3$

38. For what value(s) of k is $x^2 – kx + k$ divisible by $x – k$?
    (A) only 0
    (B) only 0 or $-\dfrac{1}{2}$

    (C) only 1
    (D) any value of k
    (E) no value of k

39. If the graphs of $x^2 = 4(y + 9)$ and $x + ky = 6$ intersect on the x-axis, then $k =$
    (A) 0
    (B) 6
    (C) –6
    (D) no real number
    (E) any real number

40. The length of the latus rectum of the hyperbola whose equation is $x^2 – 4y^2 = 16$ is

    (A) 1
    (B) 2
    (C) $\sqrt{20}$
    (D) $2\sqrt{20}$
    (E) 16

41. If $f_n = \begin{cases} \dfrac{f_{n-1}}{2} & \text{when } f_{n-1} \text{ is an even number} \\ 3 \cdot f_{n-1} + 1 & \text{when } f_{n-1} \text{ is an odd number} \end{cases}$

    and $f_1 = 3$, then $f_5 =$

    (A) 1
    (B) 2
    (C) 4
    (D) 8
    (E) 16

    [handwritten: $f_5 = \dfrac{f_{n-1}}{2}$]

    [handwritten scratch work: $f_1 = 3$, $f_2 = 10$, $f_3 = 5$, $f_4 = 16$, $f_5 = 8$]

42. How many distinguishable rearrangements of the letters in the word CONTEST start with the two vowels?

    (A) 120
    (B) 60
    (C) 10
    (D) 5
    (E) None of these

    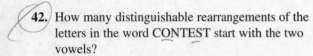
    [handwritten: 2 1 _ _ _ _ _]

43. Which of the following translations of the graph of $y = x^2$ would result in the graph of $y = x^2 - 6x + k$, where k is a constant greater than 10.

    (A) Left 6 units and up k units
    (B) Left 3 units and up $k + 9$ units
    (C) Right 3 units and up $k + 9$ units
    (D) Left 3 units and up $k - 9$ units
    (E) Right 3 units and up $k - 9$ units

    [handwritten: $\dfrac{6}{2} = 3$]

44. How many positive integers are there in the solution set of $\dfrac{x}{x-2} > 5$?

    (A) 0
    (B) 2
    (C) 4
    (D) 5
    (E) an infinite number

45. During the year 1995 the price of ABC Company stock increased by 125%, and during the year 1996 the price of the stock increased by 80%. Over the period from January 1, 1995, through December 31, 1996, by what percentage did the price of ABC Company stock rise?

    (A) 103%
    (B) 205%
    (C) 305%
    (D) 405%
    (E) 505%

    [handwritten: $\left(x(1 + 1.25) \right) 1.8$]

USE THIS SPACE FOR SCRATCH WORK

46. If $x_0 = 3$ and $x_{n+1} = x_n \sqrt{x_n + 1}$, then $x_3 =$

(A) 15.9
(B) 31.7
(C) 44.9
(D) 65.2
(E) 173.9

$x_1 = 3\sqrt{3+1} = 6$

$x_2 = 6\sqrt{6+1} =$

47. When the smaller root of the equation $3x^2 + 4x - 1 = 0$ is subtracted from the larger root, the result is

(A) −1.3
(B) 0.7
(C) 1.3
(D) 1.8
(E) 2.0

48. Each of a group of 50 students studies either French or Spanish but not both, and either math or physics but not both. If 16 students study French and math, 26 study Spanish, and 12 study physics, how many study both Spanish and physics?

(A) 4
(B) 5
(C) 6
(D) 8
(E) 10

50

49. If x, y, and z are positive, with $xy = 24$, $xz = 48$, and $yz = 72$, then $x + y + z =$

(A) 22
(B) 36
(C) 50
(D) 62
(E) 96

$x = \frac{24}{y}$

$z = \frac{72}{y}$

$z = \frac{48}{x} = \frac{48}{\frac{24}{y}} = 48 \cdot \frac{y}{24} = 2y$

$\frac{24}{y} + y + 2y =$

$\frac{24}{y} + 3y$

$2y = \frac{72}{y}$

$2y^2 = 72$

$y^2 = \frac{72}{2}$

$y = 6$

$x = 4$

$z = 12$

50. $\sin^{-1}(\cos 100°) =$

(A) −1.4
(B) −0.2
(C) 0.2
(D) 1.0
(E) 1.4

STOP

If there is still time remaining, you may review your answers.

Model Test 5

Answer Key

MODEL TEST 5

1. **E**	14. **B**	27. **D**	40. **B**
2. **C**	15. **B**	28. **B**	41. **D**
3. **D**	16. **D**	29. **D**	42. **A**
4. **A**	17. **C**	30. **D**	43. **E**
5. **A**	18. **C**	31. **D**	44. **A**
6. **E**	19. **C**	32. **C**	45. **C**
7. **D**	20. **D**	33. **B**	46. **D**
8. **D**	21. **C**	34. **D**	47. **D**
9. **B**	22. **B**	35. **D**	48. **A**
10. **B**	23. **D**	36. **E**	49. **A**
11. **C**	24. **E**	37. **C**	50. **B**
12. **A**	25. **C**	38. **A**	
13. **D**	26. **C**	39. **E**	

ANSWERS EXPLAINED

The following explanations are keyed to the review portions of this book. The number in brackets after each explanation indicates the appropriate section in the Review of Major Topics (Part 2). If a problem can be solved using algebraic techniques alone, [algebra] appears after the explanation, and no reference is given for that problem in the Self-Evaluation Chart at the end of the test.

In these solutions the following notation is used:

i: calculator unnecessary
a: calculator helpful or necessary
g: graphing calculator helpful or necessary

1. i **(E)** Factor out $x^{2/3}$, the greatest common factor of $x^{2/3} + x^{4/3}$, to get $x^{2/3} + x^{4/3} = x^{2/3}(1 + x^{2/3})$. [algebra]

2. i **(C)** When $x = 0$, y and z can be any value. Therefore, any point in the yz-plane is a possible member of the set. [2.2]

3. i **(D)** $\dfrac{\dfrac{a}{b}}{\dfrac{1}{a} - \dfrac{1}{b}} \cdot \dfrac{ab}{ab} = \dfrac{a^2}{b - a}$. [algebra]

4. g **(A)** Enter f in Y_1 and g in Y_2, return to the home screen and enter $Y_1(Y_2(3))$ to get the correct answer choice. [1.1]

5. g **(A)** Plot the graph of $y = x - \sin x$ and observe that the graph lies above the x-axis for all $x > 0$. [1.3]

6. g **(E)** Plot the graph of $y = 3x^2 - 5$ in the standard window. The symmetry about the y-axis indicates that the zeros are opposites and therefore sum to zero.

   An alternative solution is to use the fact that the zeros of a quadratic function sum to $-\dfrac{b}{a} = 0$. [1.2]

7. i **(D)** If the plane is perpendicular to the axis of the cylinder, the intersection will be a circle, so I is possible. If the plane is parallel to the axis of the cylinder, the intersection will be parallel lines, so II is possible. It is impossible for the intersection of a plane and a cylinder to form intersecting lines because there are no intersecting lines on a cylinder. [2.2]

8. i **(D)** The slope of the line through (x, y) and $(0,0) = \dfrac{y}{x}$. The slope of the line through (y, cy) and $(0,0) = \dfrac{cy}{y} = c$. The two slopes are equal, and so $c = \dfrac{y}{x}$. [1.2]

9. g **(B)** Plot f in a standard window and use TRACE to observe that y increases without bound on either side of $x = 5$.

   An alternative solution can be found by observing that the denominator of f is $(x - 5)^2$, which equals zero only when $x = 5$. [1.2]

10. g (B) Plot the function P in a standard window and observe that it crosses the x-axis only once to the right of the y-axis. Descartes' Rule of Signs guarantees at most one zero because $P(x)$ has only 1 sign change. [1.2]

11. i (C) The powers of the imaginary unit i follow a cyclical pattern: $i^3 = -i$, and $i^6 = i^2 = -1$. Thus the equation becomes $3yi - 2i = 7i - 6 + 3x$. When written as a pair of complex numbers in standard form, it becomes $0 + (3y - 2)i = (3x - 6) + 7i$. Equating the real and imaginary parts and then solving yields $x = 2$ and $y = 3$. [3.2]

12. i (A) The slope of $f(x) = \dfrac{-2-1}{4-2} = \dfrac{-3}{2}$. Using the point-slope form, $f(x) - 1 =$

$\dfrac{-3}{2}(x - 2)$. Therefore, $f(x) = \dfrac{-3}{2}x + 4$. [1.2]

13. i (D) $s = r\theta$. $2r = r\theta$. $\theta = 2^R$. [1.3]

14. g (B) Plot the graphs of $y = 2^x + 1$ in the standard window. Since $f^{-1}(7)$ is the value of x such that $f(x) = 7$, also plot the graph of $y = 7$. The correct answer choice is the point where these two graphs intersect.

An alternative solution is to solve the equation $2^x + 1 = 7$: $x = \log_2 6 = \dfrac{\log 6}{\log 2} \approx 2.6$. [1.4]

15. g (B) Evaluate the 2 by 2 determinant to get $2x^2 - x = 3$. Use the quadratic equation program to solve $2x^2 - x - 3 = 0$ to get the correct solution.

An alternative solution is to factor $2x^2 - x - 3$ into $(2x - 3)(x + 1)$ and set each factor to zero to get $x = \dfrac{3}{2}$ or $x = -1$. [3.3]

16. g (D) Use LIST/seq to construct the sequence as seq($30 - 3x, x, 0, 70$) and store this sequence in a list (e.g., L_1). On the Home Screen, enter $L_1(71)$ for the 71st term, -180.

An alternative solution is to use the formula for the nth term of an arithmetic sequence: $t_{71} = 30 - (70)(3) = -180$. [3.4]

17. a (C) Put your calculator in radian mode: $5x = \tan^{-1} 3 \approx 1.249$. $x \approx 0.2498$. Therefore, $\tan x = 0.255139 \approx 0.3$. [1.3]

18. a (C) Take $\log_{4.05}$ of both sides of the equation, getting $p = \log_{4.05} 5.25^q$, or $p = q\log_{4.05} 5.25$. Dividing both sides by q and changing to base 10 yields

$\dfrac{p}{q} = \dfrac{\log 5.25}{\log 4.05} \approx 1.19$. [1.4]

19. a (C) Area of one base $= \pi r^2 = 4\pi$.
Lateral area $= 2\pi rh = 36\pi$.
Lateral area $-$ two bases $= 36\pi - 8\pi = 28\pi \approx 87.96 \approx 88$. [2.2]

20. a (D) Put your calculator in degree mode and evaluate $\tan^{-1}(\cos 67°)$. [1.3]

21. g (C) Plot the graph of $y = x^3 + 18x - 30$ in a $[0,3]$ by $[-5,5]$ window, and use CALC/zero to locate a zero at $x \approx 1.48$. [1.2]

22. a **(B)** The angle opposite the 37 side (call it $\angle A$) is the largest angle. By the law of

cosines, $37^2 = 23^2 + 32^2 - 2(23)(32) \cos A$.

$$\text{Cos } A = \frac{37^2 - 23^2 - 32^2}{-2(23)(32)} \approx \frac{-184}{-1472} = 0.125.$$

Therefore, $A = \cos^{-1}(0.125) \approx 83°$.   [1.3]

23. i **(D)** The domain of g is $x \geq -1$ because $x + 1 \geq 0$. The domain of $f \circ g$ consists of all x such that $g(x)$ is in the domain of f. Since $g(3) = 2$ and 2 is not in the domain of f, 3 must be excluded from the domain of $f \circ g$. [1.1]

24. a **(E)** There are four 7s in a deck, and so P(first draw is a 7) = $\dfrac{4}{52} = \dfrac{1}{13}$. There are

now only three 7s among the remaining 51 cards, and so P(second draw is a 7) =

$\dfrac{3}{51} = \dfrac{1}{17}$. Therefore, P(both draws are 7s) = $\dfrac{1}{13} \cdot \dfrac{1}{17} = \dfrac{1}{221} \approx 0.00452 \approx 0.005$. [4.2]

25. a **(C)** Since $\sqrt{y} = 3.216$, $y \approx 10.34$, and $10y \approx 103.4$, so $\sqrt{10y} \approx 10.17$. [algebra]

26. g **(C)** Plot the graph of $y = \log\sqrt{2x^2 - 15}$ in the standard window and zoom in once. Use the TRACE to determine that there are no y values on the graph between the approximate x values of -2.7 and 2.7.

An alternative solution is to use the fact that the domain of the square root function is $x \geq 0$ and its range is $y \geq 0$. However, since the domain of the log function is $x > 0$, the domain of the function f must satisfy $2x^2 - 15 > 0$, or $x^2 > 7.5$. The approximate solution to this inequality is the correct answer choice C. [1.1,1.4]

27. a **(D)** Twenty percent of 400,000 women is 80,000, and 60% of 800,000 men is 480,000. Altogether 560,000 subscribers read ads, or about 47%. [4.1]

28. i **(B)** Equate S and T using the formula for the sum of the first n terms of an arithmetic

series: $\dfrac{n}{2}[6 + (n-1)4] = \dfrac{n}{2}[16 + (n-1)2]$. This equation reduces to $n^2 - 6n = 0$,

for which b is the only positive solution. [3.4]

29. g **(D)** Plot the graph of $y = \sqrt{(1 + \sin(x)^2)}$ in a $\left[-\dfrac{\pi}{4}, \dfrac{\pi}{4}\right]$ by $[-2,2]$ window, and

observe that the maximum value of y occurs at the endpoints. Return to the home screen, and enter $Y_1(\pi/4)$ to get the correct answer choice.

An alternative solution uses the fact that on the interval $\left[-\dfrac{\pi}{4}, \dfrac{\pi}{4}\right]$, the maximum value

of $|\sin x| = \dfrac{\sqrt{2}}{2}$. Therefore, the maximum value of $f(x)$ is $\sqrt{1 + \dfrac{1}{2}} = \sqrt{\dfrac{3}{2}} \approx 1.2$. [1.3]

30. a (D) $\sin \theta = \dfrac{4}{5}$.

Therefore, $\theta = \sin^{-1}\left(\dfrac{4}{5}\right) \approx 53°$. [1.3, 2.1]

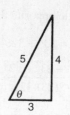

31. g (D) Plot the graph of $f(f(x))$ in the standard window by entering $Y_1 = x^2 - 4$, $Y_2 = Y_1(x)$ and de-selecting Y_1. The graph has 4 zeros located symmetrically about the *y*-axis, making answer choice D the only possible one.

An alternative solution is to evaluate $f(f(x)) = (x^2 - 4)^2 - 4 = 0$ and solve by setting $x^2 - 4 = \pm 2$, or $x^2 = 6$ or 2, so $x = \pm\sqrt{6}$ or $\pm\sqrt{2}$. Again, D is the only answer choice with 4 solutions. [1.2]

32. g (C) Enter *f* into Y_1 and *g* into Y_2. Evaluate $Y_2(Y_1(2))$ for the correct answer choice.

An alternative solution is to use the properties of logs to evaluate $f(2) = 2\log 2 = \log 4$ and $g(\log 4) = 10^{\log 4} = 4$ without a calculator. [1.4]

33. a (B) $f\left(\sqrt{2}\right) = \left(\sqrt{2}\right)^{\sqrt{2}} \approx 1.5$. [1.4]

34. a (D) The width of each rectangle is 2 and the heights are $5^0 = 1$, $5^2 = 25$, and $5^4 = 625$. Therefore, the total area is $2(1 + 25 + 625) = 1302$. [1.4]

35. g (D) Plot the graphs of $y = \pm\sqrt{9 - x^2}$ and $y = \pm\sqrt{-x^2 + 6x - 5}$ in the standard window, but with FORMAT set to GridOn. The "grid" consists exactly of the lattice points. ZOOM/ZBox around the area enclosed by the two graphs, and count the number of lattice points in that area to be 3. The points (1,0) and (3,0) appear close to the boundary, but a mental check finds that (1,0) is on the boundary of the second curve, while (3,0) is on the boundary of the first. [2.1]

36. a (E) First observe the facts that *A* is in Quadrant II and *B* is in Quadrant IV imply that $A + B$ is in Quadrant I or II, so that $\sin (A + B) > 0$. With your calculator either in degree or radian mode, enter $\sin(\sin^{-1}(3/5) + \cos^{-1}(1/3))$ to get the correct answer choice.

An alternative solution is to use the formula for the sin of a sum of two angles. Using $\sin^2 x + \cos^2 x = 1$ and the fact that *A* is in Quadrant I, together with $\sin A = \dfrac{3}{5}$, implies $\cos A = -\dfrac{4}{5}$. Similarly, $\cos B = \dfrac{1}{3}$ and *B* in Quadrant IV implies $\sin B = -\dfrac{2\sqrt{2}}{3}$.

Substituting these values into $\sin (A + B) = \sin A \cos B + \cos A \sin B$ yields the correct answer choice. [1.3]

37. i (C) $f(x) = f\left(2 \cdot \dfrac{x}{2}\right) = \left(\dfrac{x}{2}\right)^2 - \dfrac{x}{2} + 3 = \dfrac{x^2}{4} - \dfrac{x}{2} + 3$. [1.1]

38. i (A) Using the factor theorem, substitute *k* for *x* and set the result equal to zero. Then $k^2 - k^2 + k = 0$, and $k = 0$. [1.2]

39. i **(E)** If the graphs intersect on the *x*-axis, the value of *y* must be zero. Since the value of *y* is zero, it does not matter what *k* is. [1.2]

40. i **(B)** $a^2 = 16$. $b^2 = 4$. Latus rectum $= \dfrac{2b^2}{a} = 2$. [2.1]

41. i **(D)**

n	1	2	3	4	5
f_n	3	10	5	16	8

[3.4]

42. a **(A)** There are 5 consonants, CNTST, but the two Ts are indistinguishable, so there are $\dfrac{5!}{2} = 60$ ways of arranging these. There are two ways of arranging the 2 vowels in the front. Therefore, there are $2 \cdot 60 = 120$ distinguishable arrangements. [3.1]

43. i **(E)** Complete the square on $x^2 - 6x$ by adding 9. Then $x^2 - 6x + k = x^2 - 6x + 9 + k - 9 = (x - 3)^2 + (k - 9)$. This expression represents the translation of x^2 by 3 units right and $k - 9$ units up. [2.1]

44. g **(A)** Plot the graph of $y = \dfrac{x}{x-2} - 5$ in the standard window and observe the vertical

asymptote at $x = 2$. $y > \dfrac{x}{x-2} - 5$ where this graph lies above the *x*-axis, and there are no integer values of *x* in this interval.

An alternative solution is to solve the equation $\dfrac{x}{x-2} = 5$ $(x \neq 2)$: $x = 5x - 10$, or $x = \dfrac{5}{2}$.

If you test values in the intervals $(-\infty, 2)$, $\left(2, \dfrac{5}{2}\right)$, and $\left(\dfrac{5}{2}, \infty\right)$ you find that only the middle interval satisfies the inequality, but it contains no integers. [1.2]

45. a **(C)** Let the starting price of the stock be $100. During the first year a 125% increase means a $125 increase to $225. During the second year an 80% increase of the $225 stock price means a $180 increase to $405. Thus, over the 2-year period the price increased $305 from the original $100 starting price. Therefore, the price increased 305%. [algebra]

46. g **(D)** Enter 3 into your calculator. Then enter $\text{Ans }\sqrt{\text{Ans}+1}$ three times to accomplish three iterations that result in x_3 and the correct answer choice. [3.4]

47. g **(D)** Use your calculator program for the Quadratic Formula to find the roots of the equation. Then subtract the smaller root (-1.54) from the larger (0.22) and round to get the correct answer choice.

An alternative solution is to substitute the values of *a*, *b*, and *c* into the Quadratic Formula to find the algebraic solutions; then subtract the smaller from the larger to find the decimal approximation to the answer:

$$x = \frac{-4 \pm \sqrt{16+12}}{6} = \frac{-2 \pm \sqrt{7}}{3}; \quad \frac{-2+\sqrt{7}}{3} - \frac{-2-\sqrt{7}}{3} = \frac{2\sqrt{7}}{3} \approx 1.8.\ [1.2]$$

48. i **(A)**

	F	S
P	c	a
M	16	b

$a + b + c + 16 = 50$, $a + b = 26$, $a + c = 12$. Subtracting the first two equations and then the first and third gives $c = 8$, $b = 22$, and $a = 4$. Four students take both Spanish and physics. [3.1]

49. i **(A)** First evaluate the ratio $\dfrac{xy}{xz} = \dfrac{y}{z} = \dfrac{24}{48} = \dfrac{1}{2}$. Cross-multiply to get $z = 2y$, and substitute in the third equation: $yz = y(2y) = 2y^2 = 72$. Therefore, $y = 6$, $z = 12$, and $x = 4$, and $x + y + z = 22$. [1.2]

50. g **(B)** With your calculator in degree mode, evaluate $\sin^{-1}(\cos(100)) = -10°$ directly. To change to radians, return your calculator to radian mode and key in $-10°$ (using ANGLE/°). This will return -0.17.

An alternative solution is to convert $-10°$ to radians by multiplying by $\dfrac{\pi}{180°}$. [1.3]

Self-Evaluation Chart for Model Test 5

Subject Area	Questions and Review Section							Right	Number Wrong	Omitted
Algebra and Functions (25 questions)	1	3	4 1.1	6 1.2	8 1.2	9 1.2		6	0	0
	10 1.2	12 1.2	14 1.4	18 1.4	21 1.2	23 1.1		4	2	0
	25	26 1.1	31 1.2	32 1.4	33 1.4	34 1.4		4	1	1
	37 1.1	38 1.2	39 1.2	44 1.2	45	47 1.2	49 1.2	3	2	2
Trigonometry (9 questions)	5 1.3	13 1.3	17 1.3	20 1.3	22 1.3	29 1.3		6	0	0
	30 1.3	36 1.3	50 1.3					3	0	0
Coordinate and Three-Dimensional Geometry (7 questions)	2 2.2	7 2.2	19 2.2	30 2.1	35 2.1	40 2.1	43 2.1	5	2	0
Numbers and Operations (9 questions)	11 3.2	15 3.3	16 3.4	28 3.4	41 3.4	42 3.1		2	1	3
	46 3.4	48 3.1	4.9 3.1					3	0	0
Data Analysis, Statistics, and Probability (2 questions)	24 4.2	27 4.1						2	0	0
TOTALS								38	8	6

Evaluate Your Performance Model Test 5	
Rating	**Number Right**
Excellent	41–50
Very good	33–40
Above average	25–32
Average	15–24
Below average	Below 15

Calculating Your Score

Raw score R = number right $- \dfrac{1}{4}$ (number wrong), rounded = _____ 36

Approximate scaled score $S = 800 - 10(44 - R)$ = _____ 720

If $R \geq 44$, $S = 800$.

Answer Sheet

MODEL TEST 6

1 Ⓐ Ⓑ Ⓒ Ⓓ	14 Ⓐ Ⓑ Ⓒ Ⓓ	27 Ⓐ Ⓑ Ⓒ Ⓓ	40 Ⓐ Ⓑ Ⓒ Ⓓ
2 Ⓐ Ⓑ Ⓒ Ⓓ	15 Ⓐ Ⓑ Ⓒ Ⓓ	28 Ⓐ Ⓑ Ⓒ Ⓓ	41 Ⓐ Ⓑ Ⓒ Ⓓ
3 Ⓐ Ⓑ Ⓒ Ⓓ	16 Ⓐ Ⓑ Ⓒ Ⓓ	29 Ⓐ Ⓑ Ⓒ Ⓓ	42 Ⓐ Ⓑ Ⓒ Ⓓ
4 Ⓐ Ⓑ Ⓒ Ⓓ	17 Ⓐ Ⓑ Ⓒ Ⓓ	30 Ⓐ Ⓑ Ⓒ Ⓓ	43 Ⓐ Ⓑ Ⓒ Ⓓ
5 Ⓐ Ⓑ Ⓒ Ⓓ	18 Ⓐ Ⓑ Ⓒ Ⓓ	31 Ⓐ Ⓑ Ⓒ Ⓓ	44 Ⓐ Ⓑ Ⓒ Ⓓ
6 Ⓐ Ⓑ Ⓒ Ⓓ	19 Ⓐ Ⓑ Ⓒ Ⓓ	32 Ⓐ Ⓑ Ⓒ Ⓓ	45 Ⓐ Ⓑ Ⓒ Ⓓ
7 Ⓐ Ⓑ Ⓒ Ⓓ	20 Ⓐ Ⓑ Ⓒ Ⓓ	33 Ⓐ Ⓑ Ⓒ Ⓓ	46 Ⓐ Ⓑ Ⓒ Ⓓ
8 Ⓐ Ⓑ Ⓒ Ⓓ	21 Ⓐ Ⓑ Ⓒ Ⓓ	34 Ⓐ Ⓑ Ⓒ Ⓓ	47 Ⓐ Ⓑ Ⓒ Ⓓ
9 Ⓐ Ⓑ Ⓒ Ⓓ	22 Ⓐ Ⓑ Ⓒ Ⓓ	35 Ⓐ Ⓑ Ⓒ Ⓓ	48 Ⓐ Ⓑ Ⓒ Ⓓ
10 Ⓐ Ⓑ Ⓒ Ⓓ	23 Ⓐ Ⓑ Ⓒ Ⓓ	36 Ⓐ Ⓑ Ⓒ Ⓓ	49 Ⓐ Ⓑ Ⓒ Ⓓ
11 Ⓐ Ⓑ Ⓒ Ⓓ	24 Ⓐ Ⓑ Ⓒ Ⓓ	37 Ⓐ Ⓑ Ⓒ Ⓓ	50 Ⓐ Ⓑ Ⓒ Ⓓ
12 Ⓐ Ⓑ Ⓒ Ⓓ	25 Ⓐ Ⓑ Ⓒ Ⓓ	38 Ⓐ Ⓑ Ⓒ Ⓓ	
13 Ⓐ Ⓑ Ⓒ Ⓓ	26 Ⓐ Ⓑ Ⓒ Ⓓ	39 Ⓐ Ⓑ Ⓒ Ⓓ	

Model Test 6

T ear out the preceding answer sheet. Decide which is the best choice by rounding your answer when appropriate. Blacken the corresponding space on the answer sheet. When finished, check your answers with those at the end of the test. For questions that you got wrong, note the sections containing the material that you must review. Also if you do not fully understand how you arrived at some of the correct answers, you should review the appropriate sections. Finally, fill out the self-evaluation chart on page 348 in order to pinpoint the topics that give you the most difficulty.

50 questions: 1 hour

Directions: Decide which answer choice is best. If the exact numerical value is not one of the answer choices, select the closest approximation. Fill in the oval on the answer sheet that corresponds to your choice.

Notes:
(1) You will need to use a scientific or graphing calculator to answer some of the questions.
(2) You will have to decide whether to put your calculator in degree or radian mode for some problems.
(3) All figures that accompany problems are plane figures unless otherwise stated. Figures are drawn as accurately as possible to provide useful information for solving the problem, except when it is stated in a particular problem that the figure is not drawn to scale.
(4) Unless otherwise indicated, the domain of a function is the set of all real numbers for which the functional value is also a real number.

Reference Information. The following formulas are provided for your information.

Volume of a right circular cone with radius r and height h: $V = \dfrac{1}{3}\pi r^2 h$

Lateral area of a right circular cone if the base has circumference c and slant height is l:

$S = \dfrac{1}{2}cl$

Volume of a sphere of radius r: $V = \dfrac{4}{3}\pi r^3$

Surface area of a sphere of radius r: $S = 4\pi r^2$

Volume of a pyramid of base area B and height h: $V = \dfrac{1}{3}Bh$

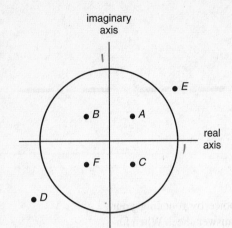

imaginary
axis

real
axis

E

B *A*

F *C*

D

USE THIS SPACE FOR SCRATCH WORK

$-2-2i$

$-2+2i$

1. In the diagram above, the circle has a radius of 1 and center at the origin. If the point *F* represents a complex number, $a + bi$, which of the other points could represent the conjugate of *F*?

    (A) A
    (B) B
    (C) C
    (D) D
    (E) E

    $a - bi$

2. For what values of *x* and *y* is $|x - y| \le |y - x|$?

    (A) $x < y$
    (B) $y < x$
    (C) $x > 0$ and $y < 0$
    (D) for no value of *x* and *y*
    (E) for all values of *x* and *y*

3. If (a,b) is a solution of the system of equations

    $\begin{cases} 2x - y = 7 \\ x + y = 8 \end{cases}$, then the difference, $a - b$, equals

    (A) −12
    (B) −10
    (C) 0
    (D) 2
    (E) 4

    $2a - b = 7 \qquad a + b = 8$

    $a + b - 1 = 7$

    $2a - b = a + b - 1$

    $a = 2b - 1$

    $a - 2b = -1$

    $\rightarrow a = 8 - b$

    $2(8 - b) - b = 7$

    $16 - 2b - b = 7$

    $16 - 3b = 7$

    $b = 3$

    $a = 5$

4. If $f(x) = x - 1$, $g(x) = 3x$, and $h(x) = \dfrac{5}{x}$, then

   $f^{-1}(g(h(5))) =$

   (A) 4
   (B) 2
   (C) $\dfrac{5}{6}$
   (D) $\dfrac{1}{2}$
   (E) $\dfrac{5}{12}$

5. A sphere is inscribed in a cube. The ratio of the volume of the sphere to the volume of the cube is

   (A) 0.79:1
   (B) 1:2
   (C) 0.52:1
   (D) 1:3.1
   (E) 0.24:1

6. If the equation $\sec^2 x - \tan x - 1 = 0$ has n solutions between $10°$ and $350°$, then $n =$

   (A) 0
   (B) 1
   (C) 2
   (D) 3
   (E) 4

7. The nature of the roots of the equation $3x^4 + 4x^3 + x - 1 = 0$ is

   (A) three positive real roots and one negative real root
   (B) three negative real roots and one positive real root
   (C) one negative real root and three complex roots
   (D) one positive real root, one negative real root, and two complex roots
   (E) two positive real roots, one negative real root, and one complex root

8. For what value(s) of k is $x^2 + 3x + k$ divisible by $x + k$?

   (A) only 0
   (B) only 0 or 2
   (C) only 0 or -4
   (D) no value of k
   (E) any value of k

USE THIS SPACE FOR SCRATCH WORK

9. What number should be added to each of the three numbers 3, 11, and 27 so that the resulting numbers form a geometric sequence?

(A) 2
(B) 3
(C) 4
(D) 5
(E) 6

$3+x \quad 11+x \quad 27+x$

$\dfrac{11+x}{3+x} = \dfrac{27+x}{11+x}$

$(11+x)^2 = (27+x)(3+x)$

$121+x^2+22x = 81+30x+x^2$

$-8x = -40$

$x = 5$

10. What is the equation of the set of points that are 5 units from point (2,3,4)?

(A) $2x + 3y + 4z = 5$
(B) $x^2 + y^2 + z^2 - 4x - 6y - 8z = 25$
(C) $(x-2)^2 + (y-3)^2 + (z-4)^2 = 25$
(D) $x^2 + y^2 + z^2 = 5$
(E) $\dfrac{x}{2} + \dfrac{y}{3} + \dfrac{z}{4} = 5$

11. If $3x^{3/2} = 4$, then $x =$

(A) 1.1
(B) 1.2
(C) 1.3
(D) 1.4
(E) 1.5

$x^{\frac{3}{2}} = \dfrac{4}{3}$

12. If $f(x) = x^3 - 4$, then the inverse of $f =$

(A) $-x^3 + 4$
(B) $\sqrt[3]{x+4}$
(C) $\sqrt[3]{x-4}$
(D) $\dfrac{1}{x^3 - 4}$
(E) $\dfrac{4}{\sqrt[3]{x}}$

$y = f^{-1}$

$x = y^3 - 4$

$\sqrt[3]{x+4}$

13. If f is an odd function and $f(a) = b$, which of the following must also be true?

I. $f(a) = -b$
II. $f(-a) = b$
III. $f(-a) = -b$ ✓

(A) only I
(B) only II
(C) only III
(D) only I and II
(E) only II and III

$f(-x) = -f(x)$

14. For all θ, $\tan\theta + \cos\theta + \tan(-\theta) + \cos(-\theta) =$

$\tan\theta + \cos\theta - \tan\theta + \cos\theta$

(A) 0
(B) $2\tan\theta$
(C) $2\cos\theta$
(D) $2(\tan\theta + \cos\theta)$
(E) 2

15. The period of the function $f(x) = k\cos kx$ is $\dfrac{\pi}{2}$. The amplitude of f is

(A) $\dfrac{1}{4}$

(B) $\dfrac{1}{2}$

(C) 1
(D) 2
(E) 4

$\dfrac{\pi}{2} = \dfrac{2\pi}{k}$

16. If $f(x) = \dfrac{x+2}{(x-2)(x^2-4)}$, its graph will have

(A) one horizontal and three vertical asymptotes
(B) one horizontal and two vertical asymptotes
(C) one horizontal and one vertical asymptote
(D) zero horizontal and one vertical asymptote
(E) zero horizontal and two vertical asymptotes

$\dfrac{x+2}{(x-2)(x+2)(x-2)} = \dfrac{1}{(x-2)^2}$

17. At a distance of 100 feet, the angle of elevation from the horizontal ground to the top of a building is 42°. The height of the building is

(A) 67 feet
(B) 74 feet
(C) 90 feet
(D) 110 feet
(E) 229 feet

$\tan 42 = \dfrac{x}{100}$

18. A sphere has a surface area of 36π. Its volume is

(A) 84
(B) 113
(C) 201
(D) 339
(E) 905

$S = 4\pi r^2$
$36\pi = 4\pi r^2$
$r = 3$

$V = \dfrac{4}{3}\pi r^3$

19. A pair of dice is tossed 10 times. What is the probability that no 7s or 11s appear as the sum of the sides facing up?

(A) 0.08
(B) 0.09
(C) 0.11
(D) 0.16
(E) 0.24

20. The lengths of two sides of a triangle are 50 inches and 63 inches. The angle opposite the 63-inch side is 66°. How many degrees are in the largest angle of the triangle?

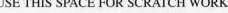

(A) 66°
(B) 67°
(C) 68°
(D) 71°
(E) 72°

$\frac{\sin 66}{63} = \frac{\sin x}{50}$

$x = 46.47$

21. Which of the following is an equation of a line that is perpendicular to $5x + 2y = 8$?

(A) $8x - 2y = 5$
(B) $5x - 2y = 8$
(C) $2x - 5y = 4$
(D) $2x + 5y = 10$

(E) $y = \dfrac{2}{-5x+8}$

$2y = -5x + 8$

$y = -\frac{5}{2}x + 4$

22. What is the period of the graph of the function

$$y = \frac{\sin x}{1 + \cos x}?$$

(A) 4π
(B) 2π
(C) π

(D) $\dfrac{\pi}{2}$

(E) $\dfrac{\pi}{4}$

$-150 \rightarrow 150$
-300

$\dfrac{300 \, \pi}{180°}$

$\bigcirc \pm 4$

$\dfrac{0 \pm \sqrt{4^2 - 4 \cdot 6 \cdot 0}}{\bigcirc}$

23. For what values of k are the roots of the equation $kx^2 + 4x + k = 0$ real and unequal?

(A) $0 < k < 2$
(B) $|k| < 2$
(C) $|k| > 2$
(D) $k > 2$
(E) $-2 < k < 0$ or $0 < k < 2$

$b^2 - 4ac > 0$
$4^2 - 4(k)(k) > 0$
$4^2 - 4k^2 > 0$

$4(4 - k^2) > 0$
$4 - k^2 > 0$
$(2+k)(2-k) > 0$
$2 + k > 0 \quad 2 - k > 0$
$k > -2 \quad -k > -2$
$\qquad\qquad k < 2$

24. A point moves in a plane so that its distance from the origin is always twice its distance from point $(1,1)$. All such points form

(A) a line
(B) a circle
(C) a parabola
(D) an ellipse
(E) a hyperbola

25. If $f(x) = 3x^2 + 24x - 53$, find the negative value of $f^{-1}(0)$.

(A) −58.8
(B) −9.8
(C) −8.2
(D) −1.8
(E) −0.2

$0 = 3x^2 + 24x - 53$

26. The operation # is defined by the equation $a \# b =$
$\dfrac{a}{b} - \dfrac{b}{a}$. What is the value of k if $3 \# k = k \# 2$?

(A) ± 2.0
(B) ± 2.4
(C) ± 3.0
(D) ± 5.5
(E) ± 6.0

$\left(\dfrac{3}{k} - \dfrac{k}{3} = \dfrac{k}{2} - \dfrac{2}{k} \right) k$

$3 - k^2 = k^2 - 2$
$5 = 2k^2$

$3 - \dfrac{k^2}{3} = \dfrac{k^2}{2} - 2$

$\dfrac{9 - k^2}{3} = \dfrac{k^2 - 4}{2}$

$2(9 - k^2) = 3(k^2 - 4)$

$18 - 2k^2 = 3k^2 - 12$

$30 = 5k^2$

$6 = k^2$

27. If $7^{x-1} = 6^x$, find x.

(A) −13.2
(B) 0.08
(C) 0.22
(D) 0.52
(E) 12.6

$\dfrac{7^x}{7} = 6^x$

$7^x = 7 \cdot 6^x$

$(x-1) \log 7 = x \log 6$

$\dfrac{x-1}{x} = .92$

$x - 1 = .92x$

$.07x - 1 = 0$

28. A red box contains eight items, of which three are defective, and a blue box contains five items, of which two are defective. An item is drawn at random from each box. What is the probability that one item is defective and one is not?

(A) $\dfrac{17}{20}$

(B) $\dfrac{5}{8}$

(C) $\dfrac{17}{32}$

(D) $\dfrac{19}{40}$

(E) $\dfrac{9}{40}$

$\left(\dfrac{3}{8} \cdot \dfrac{3}{5} \right) + \left(\dfrac{5}{8} \cdot \dfrac{2}{5} \right)$

$\dfrac{9}{40} + \dfrac{10}{40}$

29. If $(\log_3 x)(\log_5 3) = 3$, find x.

(A) 5
(B) 9
(C) 25
(D) 81
(E) 125

$\log_3 x = 4.3$

30. If $f(x) = \sqrt{x}$, $g(x) = \sqrt[3]{x+1}$, and $h(x) = \sqrt[4]{x+2}$,

then $f(g(h(2))) =$

(A) 1.2
(B) 1.4
(C) 2.9
(D) 4.7
(E) 8.5

[handwritten: $h(2) = 1.41$ $g(1,41) =$]

31. In $\triangle ABC$, $\angle A = 45°$, $\angle B = 30°$, and $b = 8$. Side $a =$

(A) 6.5
(B) 11
(C) 12
(D) 14
(E) 16

[handwritten: $\frac{\sin 30}{8} = \frac{\sin 45}{a}$]

32. The equations of the asymptotes of the graph of
$4x^2 - 9y^2 = 36$ are

(A) $y = x$ and $y = -x$
(B) $y = 0$ and $x = 0$
(C) $y = \frac{2}{3}x$ and $y = -\frac{2}{3}x$
(D) $y = \frac{3}{2}x$ and $y = -\frac{3}{2}x$
(E) $y = \frac{4}{9}x$ and $y = -\frac{4}{9}x$

[handwritten: $\frac{x^2}{9} - \frac{y^2}{4} = 1$ $\frac{2}{3}$]

33. If $g(x - 1) = x^2 + 2$, then $g(x) =$

(A) $x^2 - 2x + 3$
(B) $x^2 + 2x + 3$
(C) $x^2 - 3x + 2$
(D) $x^2 + 2$
(E) $x^2 - 2$

[handwritten: $g(x) = g(x+1-1)$ $(x+1)^2 + 2$ $x^2 + 1 + 2x + 2$]

34. If $f(x) = 3x^3 - 2x^2 + x - 2$, then $f(i) =$

(A) $-2i - 4$
(B) $4i - 4$
(C) $4i$
(D) $-2i$
(E) 0

[handwritten: $3i^3 - 2i^2 + i - 2$]

35. If the hour hand of a clock moves k radians in 48
minutes, $k =$

(A) 0.3
(B) 0.4
(C) 0.5
(D) 2.4
(E) 5

[handwritten: $\frac{60 \min}{2\pi} = \frac{48 \min}{k}$ $60k = 96\pi$ $\frac{60}{\frac{\pi}{6}} = \frac{48}{k}$]

36. If the longer diagonal of a rhombus is 10 and the large angle is 100°, what is the area of the rhombus?

(A) 37
(B) 40
(C) 42
(D) 45
(E) 50

37. Let $f(x) = \sqrt{x^3 - 4x}$ and $g(x) = 3x$. The sum of all values of x for which $f(x) = g(x)$ is

(A) −8.5
(B) 0
(C) 8
(D) 9
(E) 9.4

38. How many subsets does a set with n elements have?

(A) n^2
(B) 2^n
(C) $\binom{2n}{n}$
(D) n
(E) $n!$

39. If $f(x) = 2^{3x-5}$, find $f^{-1}(16)$.

(A) 1
(B) 2
(C) 3
(D) 4
(E) 5

40. For what positive value of n are the zeros of $P(x) = 5x^2 + nx + 12$ in ratio 2:3?

(A) 0.42
(B) 1.32
(C) 4.56
(D) 15.8
(E) 25

41. If Arcsin $x = 2$ Arccos x, then $x =$

(A) 0.9
(B) 0.5
(C) 0
(D) ±0.9
(E) ±0.5

42. A man piles 150 toothpicks in layers so that each layer has one less toothpick than the layer below. If the top layer has three toothpicks, how many layers are there?

(A) 15
(B) 17
(C) 20
(D) 148
(E) 11,322

$a_n = a_1 + (n-1)d$

$a_1 = a_1 + (n-1)d$

$0 = (n-1)d$

$150 = \frac{n}{2}(a_1 + 3)$

$150 = \frac{n}{2}($

43. If the circle $x^2 + y^2 - 2x - 6y = r^2 - 10$ is tangent to the line $12y = 60$, the value of r is

(A) 1
(B) 2
(C) 3
(D) 4
(E) 5

$y = 5$

$x^2 - 2x \underline{+1} + y^2 - 6y \underline{+9} = r^2 - 10 + 1 + 9$

$(x-1)^2 + (y-3)^2 = r^2$

44. If $a_0 = 0.4$ and $a_{n+1} = 2|a_n| - 1$, then $a_5 =$

(A) −0.6
(B) −0.2
(C) 0.2
(D) 0.4
(E) 0.6

$a_1 =$

45. If $5.21^p = 2.86^q$, what is the value of $\frac{p}{q}$?

(A) −0.60
(B) 0.55
(C) 0.60
(D) 0.64
(E) 1.57

$p \log 5.21 = q \log 2.86$

$\frac{p}{q} =$

46. As $n \to \infty$, find the limit of the product

$(\sqrt[3]{3})(\sqrt[6]{3})(\sqrt[12]{3}) \cdots (\sqrt[3 \cdot 2^n]{3})$.

(A) 1.9
(B) 2.0
(C) 2.1
(D) 2.2
(E) 2.3

$3^{\frac{3 \cdot 2^n}{2}}$

$3^{\frac{3}{2}} \times \_\_ \_\_ 3^{\frac{3 \cdot 2^n}{2}}$

$S = \dfrac{\sqrt[3]{3}}{1 - 3^{\frac{1}{2}}}$

47. There is a linear relationship between the number of chirps made by a cricket and the air temperature. A least-squares fit of data collected by a biologist yields the equation:

    estimated temperature in °F = 22.8 + (3.4)
    (the number of chirps per minute)

    What is the estimated increase in temperature that corresponds to an increase of 5 chirps per minute?

    (A) 3.4°F
    (B) 17.0°F
    (C) 22.8°F
    (D) 26.2°F
    (E) 39.8°F

48. If the length of the diameter of a circle is equal to the length of the major axis of the ellipse whose equation is $x^2 + 4y^2 - 4x + 8y - 28 = 0$, to the nearest whole number, what is the area of the circle?

    (A) 28
    (B) 64
    (C) 113
    (D) 254
    (E) 452

49. The force of the wind on a sail varies jointly as the area of the sail and the square of the wind velocity. On a sail of area 50 square yards, the force of a 15-mile-per-hour wind is 45 pounds. Find the force on the sail if the wind increases to 45 miles per hour.

    (A) 135 pounds
    (B) 225 pounds
    (C) 405 pounds
    (D) 450 pounds
    (E) 675 pounds

50. If the riser of each step in the drawing above is 6 inches and the tread is 8 inches, what is the value of |AB|?

(A) 40 inches
(B) 43.9 inches
(C) 46.6 inches
(D) 48.3 inches
(E) 50 inches

Answer Key
MODEL TEST 6

1. **B**	14. **C**	27. **E**	40. **D**
2. **E**	15. **E**	28. **D**	41. **A**
3. **D**	16. **C**	29. **E**	42. **A**
4. **A**	17. **C**	30. **A**	43. **B**
5. **C**	18. **B**	31. **B**	44. **C**
6. **D**	19. **A**	32. **C**	45. **D**
7. **D**	20. **C**	33. **B**	46. **C**
8. **B**	21. **C**	34. **D**	47. **B**
9. **D**	22. **B**	35. **B**	48. **C**
10. **C**	23. **E**	36. **C**	49. **C**
11. **B**	24. **B**	37. **E**	50. **B**
12. **B**	25. **B**	38. **B**	
13. **C**	26. **B**	39. **C**	

ANSWERS EXPLAINED

The following explanations are keyed to the review portions of this book. The number in brackets after each explanation indicates the appropriate section in the Review of Major Topics (Part 2). If a problem can be solved using algebraic techniques alone, [algebra] appears after the explanation, and no reference is given for that problem in the Self-Evaluation Chart at the end of the test.

> In these solutions the following notation is used:
>
> i: calculator unnecessary
> a: calculator helpful or necessary
> g: graphing calculator helpful or necessary

1. i **(B)** If $F = a + bi$, its conjugate $a - bi$ must be at point B. [3.2]

2. i **(E)** Since $|x - y|$ and $|y - x|$ both represent the distance between x and y, they must be equal, and they are equal for any values of x and y. [1.6]

3. i **(D)** Adding the equations gives $3x = 15$. $x = 5$ and $y = 3$. $a - b = 2$. [1.2]

4. i **(A)** $h(5) = 1$. $g(1) = 3$. Interchange x and y to find that $f^{-1}(x) = x + 1$, and so $f^{-1}(3) = 4$. [1.1]

5. a **(C)** Diameter of sphere = side of cube.

   Volume of sphere = $\dfrac{4}{3}\pi r^3$.

   Volume of cube = $s^3 = (2r)^3 = 8r^3$.

   $$\frac{\text{Volume of sphere}}{\text{Volume of cube}} = \frac{\frac{4}{3}\pi r^3}{8r^3} = \frac{\pi}{6} \approx \frac{3.14}{6} \approx \frac{0.52}{1}. \quad [2.2]$$

6. g **(D)** With your calculator in degree mode, plot the graph of $y = (1/\cos(x)^2) - \tan(x) - 1$ in a $[10,350]$ by $[-3,3]$ window and observe that the graph crosses the x-axis 3 times.

   An alternative solution is to use the identity $\sec^2 x = 1 + \tan^2 x$, so the equation becomes $\tan^2 x - \tan x = 0$. Factoring and setting each factor to zero yields $\tan x = 0$ or $\tan x - 1 = 0$. On the interval $[10°,350°]$, $\tan x = 0$ when $x = 180°$, and $\tan x = 1$ when $x = 45°$ or $225°$. [1.3]

7. g **(D)** Plot the graph of $y = 3x^4 + 4x^3 + x - 1$ in the standard window. Observe that the graph crosses the x-axis twice—once at a positive x value and once at a negative one. Since the function is a degree 4 polynomial, there are 4 roots, so the other two must be complex conjugates. [1.2]

8. i **(B)** If $x^2 + 3x + k$ is divisible by $x + k$, then $(-k)^2 + 3(-k) + k = 0$, or $k^2 - 2k = 0$. Factoring and solving yields the correct answer choice. [1.2]

9. i **(D)** If x is the number, then to have a geometric sequence, $\dfrac{11+x}{3+x} = \dfrac{27+x}{11+x}$.

   Cross-multiplying yields $121 + 22x + x^2 = 81 + 30x + x^2$, and subtracting x^2 and solving gives the desired solution.

An alternative solution is to backsolve. Determine which answer choice yields a sequence of 3 integers with a common ratio. [3.4]

10. i **(C)** The set of points represents a sphere with equation $(x-2)^2 + (y-3)^2 + (z-4)^2 = 5^2$. [2.2]

11. g **(B)** Plot the graph of $y = 3x^{3/2} - 4$ in the standard window and use CALC/zero to find the correct answer choice.

An alternative solution is to divide the equation by 3 to get $x^{3/2} = \dfrac{4}{3}$ and then raise both sides to the $\dfrac{2}{3}$ power: $x = \left(\dfrac{4}{3}\right)^{2/3} \approx 1.2$. [1.4]

12. i **(B)** Let $y = f(x) = x^3 - 4$. To get the inverse, interchange x and y and solve for y.

$x = y^3 - 4$. $y = \sqrt[3]{x+4}$. [1.1]

13. i **(C)** Use the definition of an odd function. If $f(a) = b$, then $f(-a) = -b$. Only III is true. [1.1]

14. i **(C)** Tangent is an odd function, so $\tan(-\theta) = -\tan\theta$, and cosine is an even function, so $\cos(-\theta) = \cos\theta$. Therefore, the sum in the problem is $2\cos\theta$. [1.3]

15. i **(E)** Period $= \dfrac{2\pi}{k} = \dfrac{\pi}{2}$. $k = 4$. Amplitude $= 4$. [1.3]

16. g **(C)** Plot the graph of $y = \dfrac{x+2}{(x-2)(x^2-4)}$ in the standard window and observe one vertical asymptote ($x = 2$) and one horizontal asymptote ($y = 0$). [1.2]

17. a **(C)** Tan $42° = \dfrac{x}{100}$, so $x = 100\tan 42 \approx 90$. [1.3]

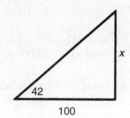

18. a **(B)** $4\pi r^2 = 36\pi$. $r^2 = 9$. $r = 3$. $V = \dfrac{4}{3}\pi r^3 = 36\pi \approx 113$. [2.2]

19. a **(A)** There are six ways to get a 7 and two ways to get an 11 on two dice, and so there are $36 - 8 = 28$ ways to get anything else. Therefore, $P(\text{no 7 or 11}) = P(\text{always getting something else}) = \left(\dfrac{28}{36}\right)^{10} \approx 0.08$. [4.2]

20. a **(C)** Use the law of sines: $\dfrac{\sin B}{50} = \dfrac{\sin 66°}{63}$. $\sin B = \dfrac{50\sin 66}{63} \approx \dfrac{45.68}{63} \approx 0.725$.

$B = \sin^{-1}(0.725) \approx 46°$ and $\angle A = 180 - 46 - 66 = 68°$. [1.3]

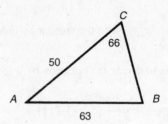

21. i **(C)** Solve the equation for y to find the slope of the line is $-\dfrac{5}{2}$. The slope of a

perpendicular line is $\dfrac{2}{5}$. Inspection of the answer choices yields C as the correct

answer. [1.2]

22. g **(B)** With your calculator in radian mode, plot the graph of $y = \sin x/(1 + \cos x)$ and
observe that the period is 2π.

An alternative solution is to use the identity $\tan\dfrac{x}{2} = \dfrac{\sin x}{1 + \cos x}$ and the fact that the

period of $\tan x$ is π to deduce the period of $\tan\dfrac{x}{2}$ as $\dfrac{\pi}{1/2} = 2\pi$. [1.3]

23. i **(E)** $b^2 - 4ac > 0$. $b^2 - 4ac = 16 - 4k^2 > 0$. $4 > k^2$. So $-2 < k < 2$. However, $k \neq 0$
because if $k = 0$, there would no longer be a quadratic equation. [1.2]

24. i **(B)** If (x,y) represents any of the points, $\sqrt{(x-0)^2 + (y-0)^2} = 2\sqrt{(x-1)^2 + (y-1)^2}$

$x^2 + y^2 = 4(x^2 - 2x + y^2 - 2y + 2)$, or $3x^2 + 3y^2 - 8x - 8y + 8 = 0$. Since the coefficients
of x^2 and y^2 are equal, the points that satisfy this equation form a circle. [2.1]

25. g **(B)** Find $f^{-1}(0)$ means to find a value of x that makes $3x^2 + 24x - 53 = 0$. Use your
program for the Quadratic Formula to determine the correct answer choice.

Use the Quadratic Formula to evaluate the solutions and then use your calculator to
find the decimal approximation to the negative solution. [1.2]

26. g **(B)** Plot the graphs of $y = \dfrac{3}{x} - \dfrac{x}{3}$ and $y = \dfrac{x}{2} - \dfrac{2}{x}$ in the standard window, and use

CALC/intersect to find the points of intersection ± 2.4.

An alternative solution is to evaluate $3\#k = \dfrac{3}{k} - \dfrac{k}{3} = \dfrac{9 - k^2}{3k}$ and $k\#2 = \dfrac{k}{2} - \dfrac{2}{k} = \dfrac{k^2 - 4}{2k}$.

Setting these two equal to each other and multiplying both sides by $6k$ yields

$18 - 2k^2 = 3k^2 - 12$, or $k^2 = 6$. Therefore, $k = \pm\sqrt{6} = \pm 2.4$. [1.1]

27. g (E) Enter $7^{(x-1)} - 6^x$ into Y_1, and generate an Auto table with TblStart = –14 and Δtbl = 1. Scan through values of x and observe a change in the sign of Y_1 between $x = 12$ and $x = 13$. Thus, E is the correct answer choice.

An alternative solution is to take $\log_7$ of both sides to get $x - 1 = \log_7 6^x = x \log_7 6 = x\dfrac{\log 6}{\log 7}$. Therefore, $x\left(1 - \dfrac{\log 6}{\log 7}\right) = 1$, so $x = \left(1 - \dfrac{\log 6}{\log 7}\right)^{-1} \approx 12.6$. [1.4]

28. i (D) Probability that an item from the red box is defective and an item from the blue box is good = $\dfrac{3}{8} \cdot \dfrac{3}{5} = \dfrac{9}{40}$. Probability that an item from the red box is good and that an item from the blue box is defective = $\dfrac{5}{8} \cdot \dfrac{2}{5} = \dfrac{10}{40}$. Since these are mutually exclusive events, the answer is $\dfrac{9}{40} + \dfrac{10}{40} = \dfrac{19}{40}$. [4.2]

29. a (E) By the change-of-base theorem, $\log_3 x = \dfrac{\log_5 x}{\log_5 3}$. Therefore $\log_5 x = 3$ and $x = 5^3 = 125$.

An alternative solution is to use the change-of-base theorem to get $\left(\dfrac{\log x}{\log 3}\right)\left(\dfrac{\log 3}{\log 5}\right) = \dfrac{\log x}{\log 5} = 3$.

$\log x = 3 \log 5 \approx 2.0969$. Therefore, $x \approx 10^{2.0969} \approx 125$. [1.4]

30. g (A) Enter f into Y_1; g into Y_2; and h into Y_3. Then return to the Home Screen and evaluate $Y_1(Y_2(Y_3(2)))$ to get the correct answer choice.

An alternative solution is to evaluate each function, starting with h in turn:

$$\sqrt{\left(\sqrt[3]{\sqrt[4]{2+2}+1}\right)} + \sqrt{\left((4)^{\wedge}(1/4)+1\right)^{\wedge}(1/3)} \approx 1.2. \ [1.1]$$

31. a (B) Use the law of sines: $\dfrac{\sin 45°}{a} = \dfrac{\sin 30°}{8}$. Therefore $a = \dfrac{8 \sin 45°}{\sin 30°} \approx 11$. [1.3]

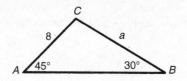

32. i (C) From this form of the equation of the hyperbola, $\dfrac{x^2}{9} - \dfrac{y^2}{4} = 1$, the equations of the asymptotes can be found from $\dfrac{x^2}{9} - \dfrac{y^2}{4} = 0$. Thus, $y = \pm\dfrac{2}{3}x$. [2.1]

33. i (B) Since $x = (x - 1 + 1)$, $g(x) = g((x - 1) + 1) = (x + 1)^2 + 2 = x^2 + 2x + 3$. [3.1]

34. g (D) Evaluate $f(i) = 3i^3 - 2i^2 + i - 2 = -2i$ on your graphing calculator. [3.2]

35. a (B) In 1 hour, the hour hand moves $\dfrac{1}{12}$ of the way around the clock, or $\dfrac{2\pi}{12} = \dfrac{\pi}{6}$

radians. $\dfrac{48}{60} \cdot \dfrac{\pi}{6} = \dfrac{2\pi}{15} \approx \dfrac{6.28}{15} \approx 0.4.$ [1.3]

36. a (C) Diagonals of a rhombus are perpendicular; they bisect each other, and they bisect

angles of the rhombus. From the figure below, $\tan 40° = \dfrac{x}{5}$, $x = 5\tan 40° \approx 4.195$.

$$A = \frac{1}{2}d_1 d_2 = \frac{1}{2}(10)(2)(4.195) \approx 42. \quad [1.3]$$

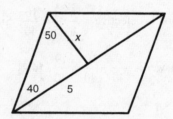

37. g (E) Plot the graphs of f and g in a $[-10, 15]$ by $[-10, 50]$ window. One point of intersection is the origin. Find the other by using CALC/intersection to get the correct answer choice.

An alternative solution is to set $\sqrt{x^3 - 4x} = 3x$, square both sides, and factor out x to get $x(x^2 - 9x - 4) = 0$. Use the Quadratic Formula with the second factor to find the solutions $x \approx -0.4, 9.4$, and observe that the first does not satisfy the original equation. Thus, the two solutions are 0 and 9.4, resulting in the correct answer choice. [1.2]

38. i (B) Each element is either in a subset or not, so there are 2 choices for each of n elements. This yields $2 \times 2 \times \cdots \times 2$ (n factors) $= 2^n$ subsets. [3.1]

39. i (C) This problem is most readily solved by substituting each answer choice into f to determine which produces the value 16. Since $f(3) = 16, f^{-1}(16) = 3$. [1.4]

40. a (D) If the zeros of $P(x)$ are in the ratio $2:3$, they must take the form $2k$ and $3k$ for some value k, and $(x - 2k)(x - 3k) = x^2 - 5k + 6k^2 = 0$. Dividing $P(x)$ by 5 and equating

coefficients yields $\dfrac{n}{5} = -5k$ and $\dfrac{12}{5} = 6k^2$. Therefore, $k = \pm\sqrt{\dfrac{2}{5}}$. Since the problem

asks for a positive value of n, we use $k = -\sqrt{\dfrac{2}{5}}$, so $n = -25k \approx 15.8$. [1.2]

41. g (A) With your calculator in radian mode, plot the graphs of $y = \sin^{-1} x$ and $y = 2\cos^{-1} x$ in a $[-2, 2]$ by $[-2, 2]$ window. Use CALC/intersection to find the correct answer choice.

An alternative solution is to let $A = \arcsin x$ and $B = \text{arc} \cos x$. Then $\sin A = \cos B = x$, so A and B are complementary, and $A = 2B$. Therefore, $B = 30°$ and $A = 60°$, and it

follows that $x = \dfrac{\sqrt{3}}{2} \approx 0.9$. [1.3]

42. i (A) This is an arithmetic series with $t_1 = 3$, $d = 1$, and $S = 150$. $150 = \dfrac{n}{2}[6 + (n-1) \cdot 1]$. $n = 15$. [3.4]

43. g **(B)** Complete the square in the equation of the circle: $(x-1)^2 + (y-3)^2 = r^2$. The center of the circle is at (1,3). If the circle is tangent to the line, then the distance between (1,3) and the line $y = 5$ is the radius of the circle, r. Therefore, $r = 2$. [2.1]

44. g **(C)** Enter 0.4 into your calculator, followed by 2|Ans| – 1 five times to get $a_5 = 0.2$.

    An alternative solution is to evaluate each a_i in turn:

    $a_1 = 2|0.4| - 1 = -0.2$
    $a_2 = 2|a_1| - 1 = 2|-0.2| - 1 = -0.6$
    $a_3 = 2|a_2| - 1 = 2|-0.6| - 1 = 0.2$
    $a_4 = 2|a_3| - 1 = 2|0.2| - 1 = -0.6$
    $a_5 = 2|a_4| - 1 = 2|-0.6| - 1 = 0.2$ [3.4]

45. a **(D)** Take the logarithms (either base 10 or base e) to get $p \log 5.21 = q \log 2.86$.

    Divide both sides of this equation by $q \log 5.21$ to get $\dfrac{p}{q} = \dfrac{\log 2.86}{\log 5.21} \approx 0.64$. [1.4]

46. g **(C)** The infinite product can be approximated by using your calculator and a "large" value of n. The exponents $\dfrac{1}{3}, \dfrac{1}{6}, \dfrac{1}{12}, \ldots$ form a geometric sequence with a first term

    of $\dfrac{1}{3}$ and constant ratio of $\dfrac{1}{2}$. Enter $prod\left(seq\left(3 \wedge \left(\left(\dfrac{1}{3}\right)\left(\dfrac{1}{2}\right) \wedge x \right), x, 0, 50\right)\right)$ to

    approximate the product of the first 50 terms as 2.08. Evaluating the product of 75 terms yields the same approximation to 9 decimals, so choose C.

    An alternative solution is to recognize that the desired product is equal to $3^{1/3+1/6+1/12+\cdots}$, so the exponent is the sum of an infinite geometric series with

    $t_1 = \dfrac{1}{3}$ and $r = \dfrac{1}{2}$. Using the formula for the sum of such a series yields $\dfrac{1/3}{1-1/2} = \dfrac{2}{3}$,

    so the desired product is $3^{2/3} \approx 2.1$. [3.4]

47. a **(B)** Temperature increases by 3.4°F for each additional chirp. Therefore, 5 additional chirps indicate an increase of 5(3.4) = 17.0°F. [1.2]

48. a **(C)** Complete the square on the ellipse formula, and put the equation in standard form:

    $x^2 - 4x + 4 + 4(y^2 + 2y - 1) = 28 + 4 + 4.$ $\dfrac{(x-2)^2}{36} + \dfrac{(y+1)^2}{9} = 1$. This leads to the

    length of the major axis: $2\sqrt{36} = 12$. Therefore, the radius of the circle is 6, and the area $= 36\pi \approx 113$. [2.1]

49. i **(C)** Since the velocity of a 45-mile-per-hour wind is 3 times that of a 15-mile-per-hour wind and the force on the sail is proportional to the square of the wind velocity, the force on the sail of a 45-mile-per-hour wind is 9 times that of a 15-mile-per-hour wind: $9 \cdot 45 = 405$. [algebra]

50. g **(B)** Total horizontal distance traveled = (4)(8) = 32. Total vertical distance traveled = (5)(6) = 30. If a coordinate system is superimposed on the diagram with A at (0,0), then B is at (32,30). Use the program on your calculator to find the distance between two points to compute the correct answer choice. [2.1]

Self-Evaluation Chart for Model Test 6

Subject Area	Questions and Review Section						Right	Number Wrong	Omitted
Algebra and Functions (22 questions)	2	3 1.2	4 1.1	7 1.2	8 1.2	11 1.4	6	0	0
	12 1.1	13 1.1	16^w 1.2	21 1.2	23 1.2	25^w 1.2	4	2	0
	26^w 1.1	27 1.4	29 1.4	30 1.1	37^w 1.2	39^w 1.4	3	3	0
	40° 1.2	45 1.4	47° 1.2	49°			1	0	3
Trigonometry (10 questions)	6 1.3	14 1.3	15 1.3	17 1.3	20 1.3	22 1.3	6	0	0
	31 1.3	35^w 1.3	36 1.3	41° 1.3			2	1	1
Coordinate and Three-Dimensional Geometry (8 questions)	5 2.2	10 2.2	18 2.2	24° 2.1	32 2.1	43^w 2.1	4	1	1
	48° 2.1	50° 2.1					0	0	2
Numbers and Operations (8 questions)	1 3.2	9 3.4	33 3.1	34 3.2	38° 3.1	42° 3.4	4	0	2
	44° 3.4	46° 3.4					0	0	2
Data Analysis, Statistics, and Probability (2 questions)	19° 4.2	28 4.2					1	0	1
TOTALS							31	7	11

Evaluate Your Performance Model Test 6	
Rating	**Number Right**
Excellent	41–50
Very good	33–40
Above average	25–32
Average	15–24
Below average	Below 15

Calculating Your Score

Raw score R = number right $- \dfrac{1}{4}$ (number wrong), rounded = _____ 29 _____

Approximate scaled score $S = 800 - 10(44 - R) =$ _____ 650 _____

If $R \geq 44$, $S = 800$.

APPENDIX

Summary of Formulas

CHAPTER 1: FUNCTIONS

1.2 Polynomial Functions

Linear Functions

General form of the equation: $Ax + By + C = 0$

Slope-intercept form: $y = mx + b$, where m represents the slope and b the y-intercept

Point-slope form: $y - y_1 = m(x - x_1)$, where m represents the slope and (x_1, y_1) are the coordinates of some point on the line

Slope: $m = \dfrac{y_1 - y_2}{x_1 - x_2}$, where (x_1, y_1) and (x_2, y_2) are the coordinates of two points

Parallel lines have equal slopes.

Perpendicular lines have slopes that are negative reciprocals.
  If m_1 and m_2 are the slopes of two perpendicular lines, $m_1 \cdot m_2 = -1$.

Distance between two points with coordinates (x_1, y_1) and $(x_2, y_2) = \sqrt{(x_1 - x_2)^2 + (y_1 - y_2)^2}$

Coordinates of the midpoint between two points $= \left(\dfrac{x_1 + x_2}{2}, \dfrac{y_1 + y_2}{2} \right)$

Distance between a point with coordinates (x_1, y_1) and a line $Ax + By + C = 0 = \dfrac{\left| Ax_1 + By_1 + C \right|}{\sqrt{A^2 + B^2}}$

If θ is the angle between two lines, $\tan \theta = \dfrac{m_1 - m_2}{1 + m_1 m_2}$, where m_1 and m_2 are the slopes of the two lines.

Quadratic Functions

General quadratic equation: $ax^2 + bx + c = 0$

General quadratic formula: $x = \dfrac{-b \pm \sqrt{b^2 - 4ac}}{2a}$

General quadratic function: $y = ax^2 + bx + c$

Coordinates of vertex: $\left(-\dfrac{b}{2a}, c - \dfrac{b^2}{4a} \right)$

Axis of symmetry equation: $x = -\dfrac{b}{2a}$

Sum of zeros (roots) $= -\dfrac{b}{a}$

Product of zeros (roots) $= \dfrac{c}{a}$

Nature of zeros (roots):
   If $b^2 - 4ac < 0$, two complex numbers
   If $b^2 - 4ac = 0$, two equal real numbers
   If $b^2 - 4ac > 0$, two unequal real numbers

1.3 Trigonometric Functions and Their Inverses

$$\sin\theta = \frac{\text{opposite}}{\text{hypotenuse}} \qquad \cos\theta = \frac{\text{adjacent}}{\text{hypotenuse}}$$

$$\tan\theta = \frac{\text{opposite}}{\text{adjacent}} \qquad \cot\theta = \frac{\text{adjacent}}{\text{opposite}}$$

$$\sec\theta = \frac{\text{hypotenuse}}{\text{adjacent}} \qquad \csc\theta = \frac{\text{hypotenuse}}{\text{opposite}}$$

$$\pi^R = 180°$$

Length of arc in circle of radius r and central angle θ is given by $r\theta^R$.

Area of sector of circle of radius r and central angle θ is given by $\dfrac{1}{2}r^2\theta^R$.

Trigonometric Reduction Formulas

1. $\sin^2 x + \cos^2 x = 1$ ⎫
2. $\tan^2 x + 1 = \sec^2 x$ ⎬ Pythagorean identities
3. $\cot^2 x + 1 = \csc^2 x$ ⎭

4. $\sin(A + B) =$
   $\sin A \cdot \cos B + \cos A \cdot \sin B$
5. $\sin(A - B) =$
   $\sin A \cdot \cos B - \cos A \cdot \sin B$
6. $\cos(A + B) =$
   $\cos A \cdot \cos B - \sin A \cdot \sin B$
7. $\cos(A - B) =$ sum and difference
   $\cos A \cdot \cos B + \sin A \cdot \sin B$ formulas
8. $\tan(A + B) =$
   $\dfrac{\tan A + \tan B}{1 - \tan A \cdot \tan B}$

9. $\tan(A - B) =$
   $\dfrac{\tan A - \tan B}{1 + \tan A \cdot \tan B}$

10. $\sin 2A = 2\sin A \cdot \cos A$

11. $\cos 2A = \cos^2 A - \sin^2 A$

12. $\qquad = 2\cos^2 A - 1$ } double-angle formulas

13. $\qquad = 1 - 2\sin^2 A$

14. $\tan 2A = \dfrac{2\tan A}{1 - \tan^2 A}$

15. $\sin \dfrac{1}{2}A = \pm\sqrt{\dfrac{1 - \cos A}{2}}$

16. $\cos \dfrac{1}{2}A = \pm\sqrt{\dfrac{1 + \cos A}{2}}$

17. $\tan \dfrac{1}{2}A = \pm\sqrt{\dfrac{1 - \cos A}{1 + \cos A}}$ } half-angle formulas

18. $\qquad = \dfrac{1 - \cos A}{\sin A}$

19. $\qquad = \dfrac{\sin A}{1 + \cos A}$

In any $\triangle ABC$:

$$\text{Law of sines}: \frac{\sin A}{a} = \frac{\sin B}{b} = \frac{\sin C}{c}$$

$$\text{Law of cosines}: \ a^2 = b^2 + c^2 - 2bc \cdot \cos A$$

$$\text{Area} = \frac{1}{2}bc \cdot \sin A$$

1.4 Exponential and Logarithmic Functions

Exponents

$$x^a \cdot x^b = x^{a+b} \qquad \frac{x^a}{x^b} = x^a - b$$

$$(x^a)^b = x^{ab}$$

$$x^0 = 1 \qquad x^{-a} = \frac{1}{x^a}$$

Logarithms

$$\log_b (p \cdot q) = \log_b p + \log_b q \qquad \log_b\left(\frac{p}{q}\right) = \log_b p - \log_b q$$

$$\log_b p^x = x \cdot \log_b p \qquad\qquad \log_b 1 = 0$$

$$\log_b p = \frac{\log_a p}{\log_a b} \qquad\qquad \log_b b = 1$$

$$b^{\log_b p} = p$$

$\text{Log}_b N = x$ if and only if $b^x = N$

1.6 Miscellaneous Functions

Absolute Value

If $x \geq 0$, then $|x| = x$.
If $x < 0$, then $|x| = -x$.

Greatest Integer Function

$[x] = i$, where i is an integer and $i \leq x < i + 1$

CHAPTER 2: GEOMETRY AND MEASUREMENT

2.1 Coordinate Geometry

Conic Sections

$$Ax^2 + Bxy + Cy^2 + Dx + Ey + F = 0$$

If $B^2 - 4AC < 0$ and $A = C$, graph is a circle.
If $B^2 - 4AC < 0$ and $A \neq C$, graph is an ellipse.
If $B^2 - 4AC = 0$, graph is a parabola.
If $B^2 - 4AC > 0$, graph is a hyperbola.

Circle

$$(x - h)^2 + (y - k)^2 = r^2$$
   with center at (h,k) and radius $= r$

Ellipse

$$\frac{(x-h)^2}{a^2} + \frac{(y-k)^2}{b^2} = 1, \text{ major axis horizontal}$$
$$\frac{(x-h)^2}{b^2} + \frac{(y-k)^2}{a^2} = 1, \text{ major axis vertical,}$$

where $a^2 = b^2 + c^2$. Coordinates of center: (h,k).

Vertices: $\pm a$ units along major axis from center

Foci: $\pm c$ units along major axis from center

Minor axis: perpendicular to major axis at center

Length $= 2b$

Eccentricity $= \dfrac{c}{a}$

Length of latus rectum $= \dfrac{2b^2}{a}$

Hyperbola

$$\frac{(x-h)^2}{a^2} - \frac{(y-k)^2}{b^2} = 1, \text{ transverse axis horizontal}$$

$$\frac{(y-k)^2}{a^2} - \frac{(x-h)^2}{b^2} = 1, \text{ transverse axis vertical, where } c^2 = a^2 + b^2. \text{ Coordinates of center: } (h,k).$$

Vertices: $\pm a$ units along the transverse axis from center

Foci: $\pm c$ units along the transverse from center

Conjugate axis: perpendicular to transverse axis at center

Eccentricity $= \dfrac{c}{a}$

Length of latus rectum $= \dfrac{2b^2}{a}$

Asymptotes: Slopes $=$

$\pm\dfrac{b}{a}$ if transverse axis is horizontal

$\pm\dfrac{a}{b}$ if transverse axis is vertical

Parabola

$(x - h)^2 = 4p(y - k)$ opens up or down—axis of symmetry is vertical

$(y - k)^2 = 4p(x - h)$, opens to the side—axis of symmetry is horizontal

Coordinates of vertex: (h,k)

Equation of axis of symmetry:
   $x = h$ if vertical
   $y = k$ if horizontal

Focus: p units along the axis of symmetry from vertex

Equation of directrix:
   $y = -p$ if axis of symmetry is vertical
   $x = -p$ if axis of symmetry is horizontal

Eccentricity $= 1$

Length of latus rectum $= 4p$

Polar Coordinates

$x = r \cdot \cos\theta \qquad y = r \cdot \sin\theta$
$x^2 + y^2 = r^2$

2.2 Three-Dimensional Geometry

Distance between two points with coordinates

(x_1,y_1,z_1) and $(x_2,y_2,z_2) = \sqrt{\left(x_1 - x_2\right)^2 + \left(y_1 - y_2\right)^2 + \left(z_1 - z_2\right)^2}$.

Distance between a point with coordinates (x_1,y_1,z_1) and a plane with equation

$Ax + By + Cz + D = 0 = \dfrac{Ax_1 + By_1 + Cz_1 + D}{\sqrt{A^2 + B^2 + C^2}}$

Triangle

$A = \dfrac{1}{2}bh$; b = base, h = height

$A = \dfrac{1}{2}ab\sin C$; a, b = any two sides, C = angle included between sides a and b

Heron's formula:

$A = \sqrt{s(s-a)(s-b)(s-c)}$; a, b, c are the three sides of the triangle, $s = \dfrac{1}{2}(a+b+c)$

Rhombus

Area $= bh = \dfrac{1}{2}d_1d_2$; b = base, h = height, d = diagonal

Cylinder

Volume $= \pi r^2 h$

Lateral surface area $= 2\pi rh$

Total surface area $= 2\pi rh + 2\pi r^2$

In all formulas, r = radius of base, h = height

Cone

Volume $= \dfrac{1}{3}\pi r^2 h$

Lateral surface area $= \pi r\sqrt{r^2 + h^2}$

Total surface area $= \pi r\sqrt{r^2 + h^2} + \pi r^2$

In all formulas, r = radius of base, h = height

Sphere

Volume $= \dfrac{4}{3}\pi r^3$

Surface area $= 4\pi r^2$

In all formulas, r = radius

CHAPTER 3: NUMBERS AND OPERATIONS

3.1 Counting

Permutations

$_nP_r = \dfrac{n!}{(n-r)!}$, where $n! = n(n-1)(n-2)\cdots 3\cdot 2\cdot 1$

Circular permutation (e.g., around a table) of n elements $= (n-1)!$

Circular permutation (e.g., beads on a bracelet) of n elements $= \dfrac{(n-1)!}{2}$

Permutations of n elements with a repetitions and with b repetitions $= \dfrac{n!}{a!\,b!}$

Combinations

$${}_nC_r = \binom{n}{r} = \dfrac{n!}{(n-r)!\,r!} = \dfrac{{}_nP_r}{n!}$$

3.2 Complex Numbers

$i^0 = 1,\ i^1 = i,\ i^{-2} = -1,\ i^3 = -i,\ i^4 = 1,\ \ldots$

$(a+bi)(a-bi) = a^2 + b^2$

3.3 Matrices

Determinants of a 2 × 2 Matrix

$$\begin{vmatrix} a & b \\ c & d \end{vmatrix} = ad - bc$$

3.4 Sequences and Series

Arithmetic Sequence (or Progression)

nth term $= t_n = t_1 + (n-1)d$

$$\text{Sum of } n \text{ terms} = S_n = \frac{n}{2}\left(t_1 + t_n\right)$$
$$= \frac{n}{2}\left[2t_1 + (n-1)d\right]$$

Geometric Sequence (or Progression)

nth term $= t_n = t_1 r^{n-1}$

$$\text{Sum of } n \text{ terms} = S_n = \frac{t_1(1 - r^n)}{1 - r}$$

If $|r| < 1,\ S_\infty = \lim_{n \to \infty} S_n = \dfrac{t_1}{1 - r}$

3.5 Vectors

If $\vec{V} = (v_1, v_2)$ and $\vec{U} = (u_1, u_2)$,

$\vec{V} + \vec{U} = (v_1 + u_1, v_2 + u_2)$

$\vec{V} \cdot \vec{U} = v_1 u_1 + v_2 u_2$

Two vectors are perpendicular if and only if $\vec{V} \cdot \vec{U} = 0$.

CHAPTER 4: DATA ANALYSIS, STATISTICS, AND PROBABILITY

4.2 Probability

$$P(\text{event}) = \frac{\text{number of ways to get a successful result}}{\text{total number of ways of getting any result}}$$

Independent events: $P(A \cap B) = P(A) \cdot P(B)$

Mutually exclusive events: $P(A \cap B) = 0$ and $P(A \cup B) = P(A) + P(B)$

Index

NOTES

NOTES

NOTES

NOTES

The following documentation applies if you purchased
SAT Math Level 2 Subject Test with CD-ROM.

Please disregard this information if your version does not contain the CD-ROM.

SYSTEM REQUIREMENTS

The program will run on a PC with:
Windows® Intel® Pentium II 450 MHz
or faster processor, 128MB of RAM
1024 × 768 display resolution
Windows 2000, XP, Vista
CD-ROM Player

The program will run on a Macintosh® with:
PowerPC® G3 500 MHz or faster
128MB of RAM
1024 × 768 display resolution
Mac OS X v.10.1 through 10.4
CD-ROM Player

INSTALLATION INSTRUCTIONS

The product will not be installed on the end-user's computer; it will run directly from the CD-ROM.
Barron's CD-ROM includes an "autorun" feature that automatically launches the application when the
CD is inserted into the CD drive. In the unlikely event that the autorun features are disabled, alternate
launching instructions are provided below.

Windows®:
Insert the CD-ROM. The program will launch automatically. If it doesn't launch automatically:
1. Click on the Start button and choose "Run."
2. Browse to the CD-ROM drive. Select the file *SATmath2.exe* and click "Open."

Macintosh®:
1. Insert the CD-ROM.
2. Double-click the CD-ROM icon.
3. Double-click *SATmath2* icon to start the program.